Instructor's Manual
for
DC/AC
Foundations of Electronics

R. Jesse Phagan
Woodstock Academy
Woodstock, Connecticut

Publisher
THE GOODHEART-WILLCOX COMPANY, INC.
Tinley Park, Illinois

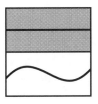

Contents

	Instructor's Manual	Textbook	Study Guide	Laboratory Manual

	Instructor's Manual	Textbook	Study Guide	Laboratory Manual

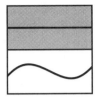

Introduction

The **DC/AC Foundations of Electronics** textbook is designed for instructional use in a two semester dc and ac electronics course. The text is divided into four areas of study: *Introductory Topics, Direct Current, Alternating Current,* and *Advanced Topics. Introductory Topics* starts the students with the basics of electricity and safety. A chapter on meter reading is also included. The *Direct Current* and *Alternating Current* chapters provide students with the laws of electrical theory and the necessary math skills. The section on *Advanced Topics* provides information on semiconductors and their applications as well as an introduction to digital circuits. An appendix, glossary, and index are also included. A Laboratory Manual and a Study Guide are available to provide hands-on activities and additional practice problems for the student.

USING THE TEXTBOOK

Each chapter in **DC/AC Foundations of Electronics** starts with a series of objectives. These are the goals that the students are expected to achieve while working through their lessons. Throughout each chapter, key words and terms are introduced. These words and terms are placed in a **bold typeface** where they first occur and are immediately defined. These new words and terms can also be found at the end of each chapter in the *Key Words and Terms Glossary*. Each chapter concludes with a summary of the important new concepts, the *Key Words and Terms Glossary,* a list of all the formulas from the chapter, and an extensive set of *Test Your Knowledge* problems to check student comprehension.

A second color is used throughout the text for emphasis. Color is used to bring attention to current, electromagnetic fields, and the doping of semiconductors. Color is used to distinguish each of the steps used to solve sample problems. Color is also used to set off each new equation. When an equation is introduced, it is set off in a color screen. Included in the screen, with each equation, is a listing of what each of the equation's variables stands for and all the proper units.

Mathematics

Electronics, as with any scientific field, involves a great deal of mathematics. It is very important for both an instructor and student to realize that it may not be in the student's best interest to try and avoid math because it is difficult. Instead, it is better to establish a method of learning the math and theory that works best for a particular class.

You must make a decision as to what is appropriate for an individual class. The chapters containing very difficult math, such as complex numbers, can be skipped without significant loss to future chapters.

DC/AC Foundations of Electronics makes every effort not to confuse electronics theory with overly difficult mathematics. This is done by explaining the theory in detail before the math is introduced. However, math is necessary to achieve a complete understanding. Sample problems in the textbook are provided to show the step-by-step process of solving problems. The sample problems are worked out methodically and in great detail. This helps prevent math errors from creeping into student solutions.

It is suggested that you review each of the sample problems with the class as a group. Students often do not study sample problems in great detail. As a result, they find it difficult to solve the assigned problems. When the steps are written out in the textbook, it is very easy for students to think the material is understood when in fact their understanding may not be complete. Each instructor must take the time to explain the problems in more detail.

When the students work out a problem, insist they write out every step, especially the formulas, every time. It is extremely important they learn a procedure to think through the steps. Otherwise, students memorize the easier Ohm's law formulas but cannot apply them when subscripts are used to represent different components in a circuit.

The Instructor

Many techniques have been experimented with to replace the instructor in the front of the classroom. Most methods work for only a small portion of the students. There is no instructional method that can reach every student without the personal touch of an instructor.

A teaching method that works well for a subject as difficult as electronics is to break each lesson into small chunks. First, introduce the new technical terms. Show how they will relate to the chapter. If possible, include some future applications for the topic. The students should spend some class time working on assignments related to the technical terms.

Next, introduce a lab exercise or some other group activity that closely relates to the topic of the day. Students learn in groups. It also helps to use the *language of electronics*.

Then examine sample problems with the class as a group. Assign problems for students to work on during class time. It is better for students to have immediate help with a problem than to attempt it as homework and experience frustration.

USING THE LABORATORY MANUAL AND STUDY GUIDE

These extra books provide student assignments and activities. It is through participation that student's learn the best. If you can get the students involved in the learning process, there will be much better communication and more successful learning.

These books are an extremely valuable asset to the student. It is a record of the student's progress and a valuable reference for the student to use for future experiences. You will also find the write-in books an excellent source of student achievement.

Laboratory Manual

The Laboratory Manual contains activities that are closely tied to the textbook. Each laboratory reinforces the important concepts in the corresponding chapter from the textbook.

The laboratories begin with objectives and a list of all the equipment that is needed to complete the activity. Following the list of equipment are step-by-step instructions that take the student through the laboratory. In these laboratories, students are expected to set up electrical circuits, collect data, and answer questions. Charts are provided in many of the activities for the data.

The materials required to perform the experiments should be readily available in your electronics laboratory. The component values are usually flexible. This makes the circuits easy to construct and easy to get working as intended. A list of all the equipment required for the Laboratory Manual is provided at the beginning of the manual.

Remind your students that working with electricity can be dangerous. Enforce safe laboratory practices. Discuss any student safety questions before beginning each laboratory.

Study Guide

The Study Guide is designed to provide a review of the essential material in each chapter for the students. The review is designed with questions that require students to *dig* through the chapter several times.

Questions in the Study Guide fall into three important groups: technical terms, electrical principles, and mathematics. The technical terms questions require students to match the new words for each chapter with their definitions. The questions on understanding electrical principles are generally multiple choice questions, with some fill-in the blank problems. All of the math questions require the students to write out the formulas and substitutions in addition to the answers.

USING THE INSTRUCTOR'S MANUAL

This Instructor's Manual provides the answers to the *Test Your Knowledge* questions in the textbook and the answers to the Study Guide questions. In addition, reproducible test masters for each chapter of the textbook are included at the end of each section. The answers to these tests are included with the answers to the other materials. Due to the wide range of component values that can be used in the activities in the Laboratory Manual, no *right* answers can be provided in this Instructor's Manual for the laboratories. However, by performing the laboratories prior to presenting the materials to the students, you can create your own answer key.

In many of the problems in the Study Guide and in the chapter tests, students are asked to supply all formulas and substitutions used to find the answer. All of these formulas and substitutions are included with the answers. These formulas and substitutions have been included to help save you time, and to help diagnose student difficulties if they answer the questions incorrectly.

ELECTRONIC INDUSTRIES FOUNDATION STANDARDS

The following chart correlates the Electronics Industries Foundation (EIF) (the not-for-profit foundation of the Electronic Industries Association) skill standards to the DC/AC Foundations of Electronics text. The skill standards are listed by their EIF Technical Skills numbers. They are correlated to the text by page number and to the Laboratory Manual (LM) by the number of the laboratory activity. Although items not covered in the text have been omitted, the Technical Skills numbers remain the same to facilitate correlation to your program. This chart allows you to gear your curriculum to the applied academic and workplace skills identified by the Electronic Industries Foundation.

SKILL STANDARDS

General		*DC/AC Correlation*
A.01	Demonstrate an understanding of proper safety techniques for all types of circuits and components (dc circuits, ac circuits, analog circuits, digital circuits, discrete solid-state circuits, microprocessors).	25–35
A.03	Demonstrate an understanding of proper trouble-shooting techniques.	168–171, 201–203, 491–495
A.05	Demonstrate an understanding of acceptable soldering/desoldering techniques, including through-hole and surface mount devices.	32–34
A.07	Demonstrate an understanding of use of data books and cross reference/technical manuals to specify and requisition electronic components.	LM 12-3, 13-3
A.08	Demonstrate an understanding of the interpretation and creation of electronic schematics, technical drawings, and flow diagrams.	Covered throughout text
A.09	Demonstrate an understanding of design curves, tables, graphs, and recording of data.	Covered throughout text
A.10	Demonstrate an understanding of color codes and other component descriptors.	29–30, 113–135

DC Circuits		*DC/AC Correlation*
B.01	Demonstrate an understanding of sources of electricity in dc circuits.	283–309
B.02	Demonstrate an understanding of principles and operation of batteries.	295–306
B.03	Demonstrate an understanding of the meaning of and relationships among and between voltage, current, resistance, and power in dc.	75–108

AC Circuits   *DC/AC Correlation*

Discrete Solid State Devices   *DC/AC correlation*

The Electronics Industry

OBJECTIVES

After studying this chapter, students should be able to:
- Write job descriptions for workers in several areas of the electronics industry.
- State specific skills and training requirements for technicians and engineers.

INSTRUCTIONAL MATERIALS

Text: Pages 11–24
　　　　Test Your Knowledge Questions, Page 24
Study Guide: Page 5
Laboratory Manual: Page 13

ANSWERS TO TEXTBOOK

Test Your Knowledge, Page 24

1. Electronic equipment manufacturing.
   Scientific research and development.
   Service and repair of equipment.
2. Student lists will vary (six responses required).
3. Computer programmer.
4. Follow written instructions for test procedures.
   Use test equipment to ensure proper operation.
   Identify defective products.
5. Use test equipment for extensive testing.
   Use accurate measurement tools.
   Develop new testing procedure.

ANSWERS TO STUDY GUIDE

Page 5

1. Any two of the following:
   Use drawing, written instructions, and visual models.
   Place parts on circuit boards.
   Assemble mechanical components.
   Use hand tools.
2. a. Design a new product.
   b. Build a prototype.
3. correct defects in equipment rejected by the testing department.
4. a licensed electrician
5. Devise tests, monitor testing.
6. repair and service technician.
7. Electrician.
8. Field service technician.
9. Associate's degree, bachelor's degree.
10. prototype.

ANSWERS TO CHAPTER TEST IN THE INSTRUCTOR'S MANUAL

Pages 17–18

1. Select any three: televisions, video equipment, VCR, stereos, radios, personal computers, computer games, burglar alarms, etc.
2. Select any three: copy machines, electronic typewriters, personal computers, fax machines, telephone equipment, etc.
3. Select any three: computers, consumer products, manufacturing, medical, communications, space exploration, national defense, transportation, etc.
4. applications (also consultant)

5. Assembly worker
    Tester
    Repair technician
6. a. the engineer: Designs a product.
    b. the technician: Builds and tests the product.
7. electrician's helper
8. electrician
9. field service technician
10. prototype

 **Chapter Test**

1

The Electronics Industry

1. List three examples of consumer products an electronic technician would be expected to repair.

   _____

   _____

   _____

2. List three examples of products an office equipment technician would be expected to repair.

   _____

   _____

   _____

3. List three industries that provide career opportunities for electronics technicians and engineers.

   _____

   _____

   _____

4. An engineer who works with the marketing department to find ways in which a product can be sold is a(n)

   _____ engineer.

5. List three types of jobs in the electronics industry that require on-the-job training, rather than formal education.

   _____

   _____

   _____

6. In a research and development department, what are the job duties of:

   a. the engineer? _____

   b. the technician? _____

7. An unlicensed person working under the direct supervision of a licensed electrician is called a(n) _____

   _____.

8. The person who is responsible wiring a house has a job title of _____.

9. A technician required to do extensive travel and to work with a minimum of supervision has a job title of

   _____ _____ _____.

10. When a design engineer builds a working model of a product, it is called a(n) _____.

OBJECTIVES

After studying this chapter, students should be able to:
- Describe the effects of electrical shock.
- Explain the factors that influence the severity of electrical shock.
- Recognize ways in which to prevent electrical shock.
- Make simple repairs to a plug, outlet, and light socket.
- Demonstrate safe soldering techniques by making solder connections.
- Define technical words used in conjunction with electrical safety.

INSTRUCTIONAL MATERIALS

Text: Pages 25–36
        Test Your Knowledge Questions, Page 36
Study Guide: Pages 7–10
Laboratory Manual: Pages 15–22

ANSWERS TO TEXTBOOK

Test Your Knowledge, Page 36

1. The effects of electricity passing through the body could include slight muscle jerks, damage to organs, or death.
2. Voltage.
   Current.
   Resistance.
3. Amount of electricity.
   Path through the body.
   Length of time of exposure.
4. No-current-path protection.
   Circuit interruption.
5. Black.
   White.
   Bare (green).
6. Black—hot wire.
   White—neutral wire.
   Bare—ground wire.
7. Safety glasses.
8. a. Make a good mechanical connection.
   b. Heat both surfaces.
   c. Apply solder.
9. Reheat the joint and apply another drop of solder.
10. Student lists of safety rules will vary.

ANSWERS TO STUDY GUIDE

Pages 7–10

1. q.
2. f.
3. r.
4. c.
5. p.
6. b.
7. j.
8. k.
9. d.
10. e.
11. n.
12. g.
13. i.
14. m.
15. h.
16. o.
17. l.
18. a.
19. b.
20. b.
21. c.
22. d.

23. a.
24. b.
25. a.
26. a.
27. a.
28. b.
29. b.
30. b.
31. c.
32. c.
33. c.
34. b.
35. b.
36. c.
37. d.
38. a.

ANSWERS TO CHAPTER TEST IN THE INSTRUCTOR'S MANUAL

Pages 21–22

1. The potential energy of an electrical source.
2. The flow of electricity.
3. The opposition a current path offers to the flow of electricity.
4. A device that prevents too much current from flowing in a circuit.
5. Hot wire.
6. Neutral wire.
7. c.
8. c.
9. d.
10. a.
11. d.
12. c.
13. Black.
14. Brass.
15. Carry current and voltage.
16. White.
17. Silver.
18. Carry current but not voltage.
19. Green or bare.
20. Green.

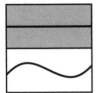

Chapter Test
2

Electrical Safety

Name: _____

Date: _____

Class: _____

Define the following technical terms.

1. Voltage: _____

2. Current: _____

3. Resistance: _____

4. Overcurrent protection device: _____

In house wiring, what is the name of the wire that best fits the description?

5. _____ The black wire, used to carry current and voltage. Attaches to the brass screw.

6. _____ The white wire, used as a current carrier, but has no voltage. Attaches to the silver screw.

Select the best answer.

_____ 7. Electric shock is possible because the body:
    a. will trip the circuit breaker.
    b. has good insulation.
    c. is an electrical conductor.
    d. has a very high resistance.

_____ 8. Which situation is most likely to have enough current to cause a fatal electric shock?
    a. Touching a ground fault interrupter while in the bathtub.
    b. Touching a car battery with a screwdriver.
    c. Touching a live wire while standing in wet grass.
    d. Touching a live wire while standing on a plastic box.

_____ 9. When a person is receiving a shock, it is best for the rescuer to:
    a. grab the person as fast as possible and pull him away.
    b. wait until the circuit breaker trips.
    c. call a rescue team by dialing 911.
    d. push the person free of the wires using a dry board.

_____ 10. A ground fault interrupter (GFI):
    a. is a circuit breaker designed to prevent a shock.
    b. is the green or bare wire in the circuit breaker panel.
    c. holds the wires at ground potential.
    d. is used in automobiles on the negative wire.

_____   11. When soldering, the tip of the soldering iron should:
        a. melt the solder prior to touching the components.
        b. be cooled on a damp sponge to prevent damage.
        c. heat one surface, then the other.
        d. make contact with both surfaces at the same time.

_____   12. A cold solder joint:
        a. will heat up when current flows through it.
        b. is the ideal solder connection.
        c. is a poor connection.
        d. looks shiny and smooth.

Complete the chart with answers that apply to standard house wiring.

	Wire Color	Screw Color	Function
Hot	13. _____	14. _____	15. _____
Neutral	16. _____	17. _____	18. _____
Ground	19. _____	20. _____	

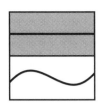

Reading Meter and Oscilloscope Scales

OBJECTIVES

After studying this chapter, students should be able to:
- State the value of numbers written with engineering notation.
- Convert numbers containing engineering notation.
- Identify digital and analog multimeters.
- Interpret the value of each line on an analog scale.
- Analyze the readings on an ohmmeter scale.
- Distinguish between the ac and dc arcs on a multimeter.
- Adjust the decimal place on a multimeter scale for each range.
- Read an oscilloscope screen for dc voltages.

INSTRUCTIONAL MATERIALS

Text: Pages 37–73
Test Your Knowledge Questions, Pages 66–73
Study Guide: Pages 11–23
Laboratory Manual: Pages 23–30

ANSWERS TO TEXTBOOK

Test Your Knowledge, Pages 66–73
1. a. 4.9×10^3
    b. 6.333333×10^6
    c. 1.023×10^{-4}
    d. -5.05×10^{-2}
    e. 1×10^1

2. a. 68,000 ohms = 68 kΩ and 0.068 MΩ
    b. 2100 watts = 2,100,000 mW and 2.1 kW
    c. 0.0039 volts = 3.9 mV and 3900 μV
    d. 0.000 042 amps = 42 μA and 42,000 nA
    e. 490 millihenrys = 0.49 H and 490,000 μH
    f. 6100 milliwatts = 0.0061 kW and 6.1 W
    g. 23,000 kilohertz = 23,000,000 Hz and 23 MHz
    h. 0.0012 kilovolts = 1,200 mV and 1.2 V
    i. 360 microfarads = 0.36 mF and 360,000 nF
    j. 6400 megohms = 6,400,000 kΩ and 6.4 GΩ

3. a. range: 10
    b. arc: DCV
    c. scale: 2, 4, 6, 8, 10
    d. each line: 0.2
    e. needle #1: 0 V dc
    f. needle #2: 5 V dc

4. a. range: 250
    b. arc: DCV,A
    c. scale: 50, 100, 150, 200, 250
    d. each line: 5
    e. needle #1: 42 mA dc
    f. needle #2: 238 mA dc

5. a. range: 50
    b. arc: ACV
    c. scale: 10, 20, 30, 40, 50
    d. each line: 1
    e. needle #1: 2 V ac
    f. needle #2: 42.2 V ac

6. a. range: 2.5
    b. arc: DCV
    c. scale: 0.5, 1.0, 1.5, 2.0, 2.5
    d. each line: 0.05
    e. needle #1: 0.75 V dc
    f. needle #2: 2.0 V dc

7. a. range: 1000
    b. arc: ACV
    c. scale: 200, 400, 600, 800, 1000
    d. each line: 20
    e. needle #1: 460 V ac
    f. needle #2: 950 V ac

8. a. range: 250
    b. arc: DCV
    c. scale: 50, 100, 150, 200, 250
    d. each line: 5
    e. needle #1: 50 V dc
    f. needle #2: 200 V dc
9. a. range: 0.5
    b. arc: DCV
    c. scale: 0.1, 0.2, 0.3, 0.4, 0.5
    d. each line: 0.01
    e. needle #1: 0.2 V dc
    f. needle #2: 0.47 V dc
10. a. range: ×1
    b. arc: Ω
    c. needle #1: 70 Ω
    d. needle #2: 7 Ω
11. a. range: ×10k
    b. arc: Ω
    c. needle #1: 360 kΩ
    d. needle #2: 85 kΩ
12. a. volts per division: 50 mV
    b. # of divisions: 3.8
    c. measurement: 190 mV
13. a. volts per division: 0.2 V
    b. # of divisions: -2.3
    c. measurement: -0.46 V
14. a. volts per division: 0.1 V
    b. # of divisions: 4.5
    c. measurement: 0.45 V
15. a. volts per division: 5 V
    b. # of divisions: 5.6
    c. measurement: 28 V

ANSWERS TO STUDY GUIDE

Pages 11–23

1. d.
2. j.
3. c.
4. b.
5. a.
6. e.
7. h.
8. g.
9. f.
10. i.
11. 68,000 = 68 kΩ and 0.068 MΩ
12. 2100 watts = 2,100,000 mW and 2.1 kW

13. 0.0039 volts = 3.9 mV and 3900 µV
14. 0.000 042 amps = 42 µA and 42,000 nA
15. 490 millihenrys = 0.49 H and 490,000 µH
16. 6100 milliwatts = 0.0061 kW and 6.1 W
17. 23,000 kilohertz = 23,000,000 Hz and 23 MHz
18. 0.0012 kilovolts = 1200 mV and 1.2 V
19. 360 microfarads = 0.36 mF and 360,000 nF
20. 6400 megohms = 6,400,000 kΩ and 6.4 GΩ
21. a. range: 10
    b. arc: ACV
    c. scale: 2, 4, 6, 8, 10
    d. each line: 0.2
    e. needle #1: 2.6 V ac
    f. needle #2: 7.2 V ac
22. a. range: 2.5
    b. arc: DCmA
    c. scale: 0.5, 1.0, 1.5, 2.0, 2.5
    d. each line: 0.05
    e. needle #1: 0.3 mA dc
    f. needle #2: 1.66 mA dc
23. a. range: 50
    b. arc: DCV
    c. scale: 10, 20, 30, 40, 50
    d. each line: 1
    e. needle #1: 2 V dc
    f. needle #2: 27.3 V dc
24. a. range: 500
    b. arc: ACV
    c. scale: 100, 200, 300, 400, 500
    d. each line: 10
    e. needle #1: 60 V ac
    f. needle #2: 270 V ac
25. a. range: 1000
    b. arc: DCV
    c. scale: 200, 400, 600, 800, 1000
    d. each line: 20
    e. needle #1: 250 V dc
    f. needle #2: 800 V dc
26. a. range: 250
    b. arc: ACV
    c. scale: 50, 100, 150, 200, 250
    d. each line: 5
    e. needle #1: 87 V ac
    f. needle #2: 169 V ac
27. a. range: 25
    b. arc: DCmA
    c. scale: 5, 10, 15, 20, 25
    d. each line: 0.5
    e. needle #1: 5 mA dc
    f. needle #2: 16.4 mA dc

28. a. range: ×1k
     b. arc: Ω
     c. needle #1: infinity
     d. needle #2: 22 kΩ
29. a. range: ×10
     b. arc: Ω
     c. needle #1: 650 Ω
     d. needle #2: 70 Ω
30. a. volts per division: 20
     b. number of divisions: 3.1
     c. measurement: 62 volts
31. a. volts per division: 20 mV
     b. number of divisions: 6.3
     c. measurement: 126 mV
32. a. volts per division: 0.1
     b. number of divisions: 5.0
     c. measurement: 0.5 volts
33. a. volts per division: 0.5
     b. number of divisions: 3.0
     c. measurement: 1.5 volts
34. a. volts per division: 2
     b. number of divisions: 2.8
     c. measurement: 5.6 volts
35. a. volts per division: 10
     b. number of divisions: –3.6
     c. measurement: –36 volts

ANSWERS TO CHAPTER TEST IN THE INSTRUCTOR'S MANUAL

Pages 27–35

1. Meter that has a needle that moves along a scale.
2. Meter that has a digital display with numbers that change with the measurement.
3. Stands for exponential notation. Writes numbers with a power of 10 as E followed by two digits.
4. A system in mathematics that uses multiplier names and powers of 10 to move the decimal point and label quantities.
5. A mathematic process of moving the decimal point and multiplying by 10 raised to a power to maintain an equivalent value.
6. 1,500,000 W and 1.5 kW
7. 2.6 mV and 2600 μV
8. 0.580 H and 580,000 μH

9.   3500 mV and 3.5 V
10. 0.980 mF and 980,000 nF
11. a. range = 25 mA
     b. arc = DCV
     c. scale: 0, 5, 10, 15, 20, 25
     d. each line = 0.5
     e. needle #1 = 6.25 mA
     f. needle #2 = 18.0 mA
12. a. range = 10 V ac
     b. arc = AC
     c. scale: 0, 2, 4, 6, 8, 10
     d. each line = 0.2
     e. needle #1 = 1.28 V
     f. needle #2 = 6.8 V
13. a. range = 500 V ac
     b. arc = ACV
     c. scale: 0, 100, 200, 300, 400, 500
     d. each line = 10
     e. needle #1 = 22 V
     f. needle #2 = 278 V
14. a. range = ×10 Ω
     b. arc = ohms
     c. needle #1 = 80 × 10 = 800 Ω
     d. needle #2 = 10.3 × 10 = 103 Ω
15. a. range = 1000 V dc
     b. arc = DC
     c. scale: 0, 200, 400, 600, 800, 1000
     d. each line = 20
     e. needle #1 = 240 V
     f. needle #2 = 740 V
16. a. volts per division = 0.5 V
     b. number of divisions = 5.0
     c. voltage = 2.5 V
17. a. volts per division = 2 V
     b. number of divisions = 3.0
     c. voltage = 6.0 V
18. a. volts per division = 5 V
     b. number of divisions = 2.8
     c. voltage = 14.0 V
19. a. volts per division = 0.05 V
     b. number of divisions = -3.6
     c. voltage = -0.18 V
20. a. volts per division = 50 mV
     b. number of divisions = 6.3
     c. voltage = 315 mV

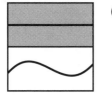

Chapter Test

3

Name: _____

Date: _____

Class: _____

Reading Meter and Oscilloscope Scales

Define the following technical terms.

1. Analog meter: _____

   _____

2. Digital meter: _____

   _____

3. E-notation: _____

   _____

4. Engineering notation: _____

   _____

5. Scientific notation: _____

   _____

Convert the number on the left to the engineering notation units shown.

Given	To Find	
6. 1500 watts	_____mW	_____kW
7. 0.0026 volts	_____mV	_____µV
8. 580 millihenrys	_____H	_____µH
9. 0.0035 kilovolts	_____mV	_____V
10. 980 microfarads	_____mF	_____nF

27

11. Use the multimeter shown below to answer the following.

_____   a. range.

_____   b. arc (ACV, DCV, Ω).

0 _____ _____ _____ _____ _____   c. numbers of scale after adjusting the decimal.

_____   d. value of each line.

_____   e. reading of needle #1.

_____   f. reading of needle #2.

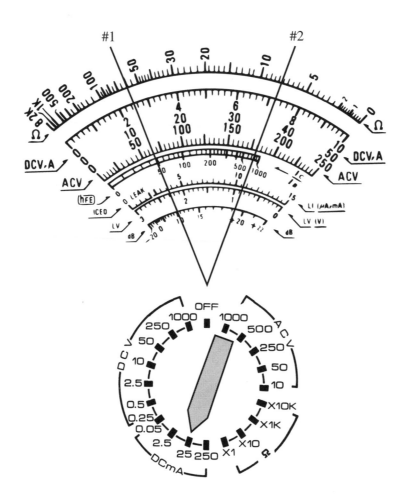

12. Use the multimeter shown below to answer the following.

_____ a. range.

_____ b. arc (ACV, DCV, Ω).

0 _____ _____ _____ _____ _____ c. numbers of scale after adjusting the decimal.

_____ d. value of each line.

_____ e. reading of needle #1.

_____ f. reading of needle #2.

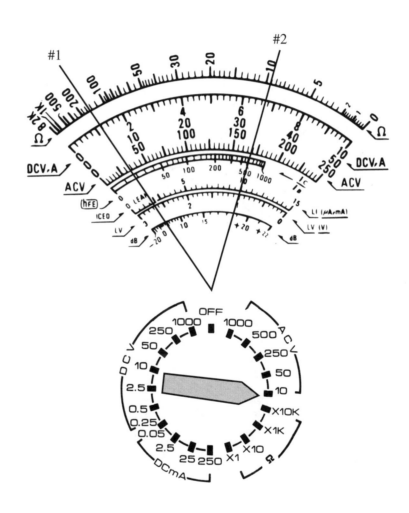

13. Use the multimeter shown below to answer the following.

_____ a. range.

_____ b. arc (ACV, DCV, Ω).

0 _____ _____ _____ _____ _____   c. numbers of scale after adjusting the decimal.

_____ d. value of each line.

_____ e. reading of needle #1.

_____ f. reading of needle #2.

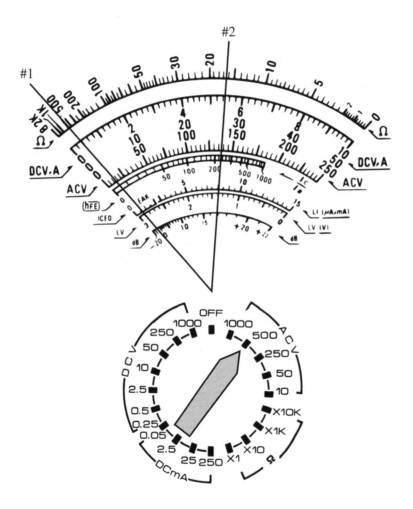

14. Use the multimeter shown below to answer the following.

_____ a. range.

_____ b. arc (ACV, DCV, Ω).

_____ c. reading of needle #1.

_____ d. reading of needle #2.

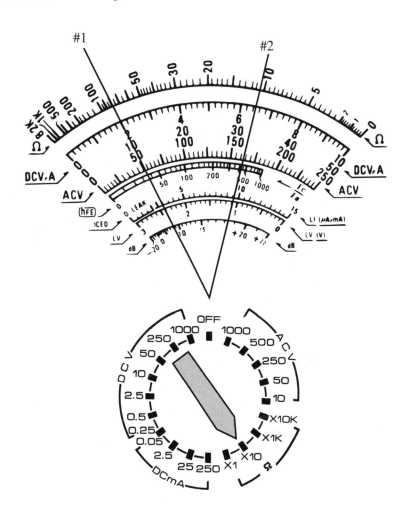

15. Use the multimeter shown below to answer the following.

_____ a. range.

_____ b. arc (ACV, DCV, Ω).

0 _____ _____ _____ _____ _____ c. numbers of scale after adjusting the decimal.

_____ d. value of each line.

_____ e. reading of needle #1.

_____ f. reading of needle #2.

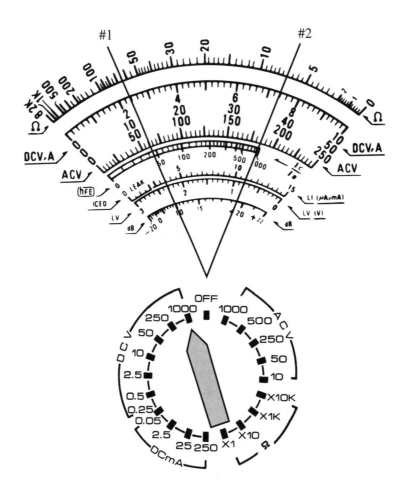

Name: _____

16. Use the oscilloscope shown below to answer the following.

    a. _____Volts per division for the selected channel.

    b. _____Number of divisions from ground to the trace. Be accurate to one decimal place.

    c. _____Value of dc voltage being measured.

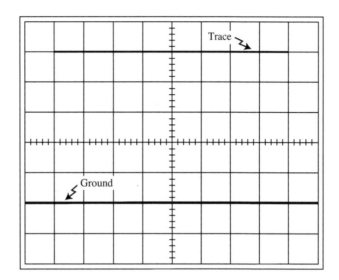

Time/Division

_____100 ms/Div_____

Channel 1
Volts/Division

_____not used_____

Channel 2
Volts/Division

_____0.5 V/Div_____

17. Use the oscilloscope shown below to answer the following.

    a. _____Volts per division for the selected channel.

    b. _____Number of divisions from ground to the trace. Be accurate to one decimal place.

    c. _____Value of dc voltage being measured.

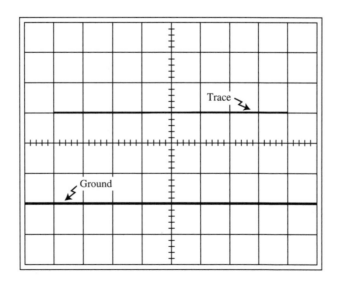

Time/Division

_____100 ms/Div_____

Channel 1
Volts/Division

_____2 V/Div_____

Channel 2
Volts/Division

_____not used_____

18. Use the oscilloscope shown below to answer the following.

    a. _____Volts per division for the selected channel.

    b. _____Number of divisions from ground to the trace. Be accurate to one decimal place.

    c. _____Value of dc voltage being measured.

Time/Division

    5 ms/Div

Channel 1
Volts/Division

    5 V/Div

Channel 2
Volts/Division

    not used

19. Use the oscilloscope shown below to answer the following.

    a. _____Volts per division for the selected channel.

    b. _____Number of divisions from ground to the trace. Be accurate to one decimal place.

    c. _____Value of dc voltage being measured.

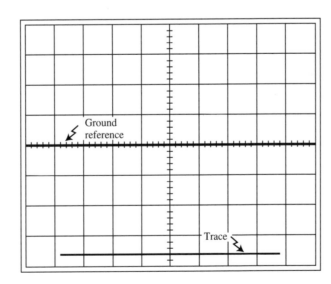

Time/Division

    500 μs/Div

Channel 1
Volts/Division

    0.05 V/Div

Channel 2
Volts/Division

    not used

20. Use the oscilloscope shown below to answer the following.

    a. _____Volts per division for the selected channel.

    b. _____Number of divisions from ground to the trace. Be accurate to one decimal place.

    c. _____Value of dc voltage being measured.

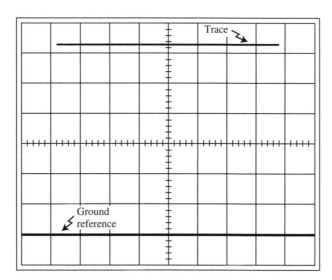

Time/Division

_____0.5 ms/Div_____

Channel 1
Volts/Division

_____50 mV/Div_____

Channel 2
Volts/Division

_____off_____

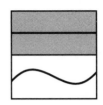

Voltage, Current, Resistance, and Power

OBJECTIVES

After studying this chapter, students should be able to:
- Describe the role of the electron in electricity.
- Define the terms: voltage, current, resistance, and power.
- Identify the units of measure and their symbols for the four basic electrical terms.
- Use engineering notation as a mathematical tool.
- Analyze the Ohm's law and power formulas in terms of direct and inverse relationships.
- Solve problems using the Ohm's law and power formulas.

INSTRUCTIONAL MATERIALS

Text: Pages 75–108
　　　　Test Your Knowledge Questions, Page 108
Study Guide: Pages 25–34
Laboratory Manual: Pages 31–40

ANSWERS TO TEXTBOOK

Test Your Knowledge, Page 108

1. Voltage must be present.
   A complete path must be available.
2. $27,030 = 27,000$
   $5.715 = 5.72$
   $0.03419802 = 0.0342$
   $0.0036226 = 0.00362$
   $100,219.3 = 100,000$
3. $V = 100$ volts
4. $I = 0.03$ amps $= 30$ mA
5. $R = 120$ ohms
6. $P = 2.4$ watts
7. $I = 0.5$ amps
8. $V = 200$ volts
9. $P = 90$ watts
10. $R = 0.4$ ohms
11. $I = 2$ amps
12. $P = 4$ watts
13. $R = 240$ ohms
14. $V = 20$ volts
15. cost $= \$0.30$
16. $R_T = 527.8$ kilohms
17. $V = 90.55$ millivolts
18. $V = 18$ volts
19. $I = 10$ milliamps
20. $P = 0.1$ watts

ANSWERS TO STUDY GUIDE

Pages 25–34

1. c.
2. u.
3. j.
4. r.
5. d.
6. m.
7. l.
8. y.
9. n.
10. e.
11. p.
12. z.
13. i.
14. s.
15. f.
16. k.
17. v.
18. t.
19. b.
20. g.
21. a.
22. x.
23. w.
24. h.
25. o.
26. q.

27. d.
28. c.
29. b.
30. e.
31. a.

32.

Electrical Quantity		Unit of Measure	
Name	**Symbol**	**Name**	**Symbol**
voltage	V	volts	V
power	P	watts	W
current	I	ampere	A
resistance	R	ohms	Ω

33. $78{,}056 = 78{,}100$
$0.151802 = 0.152$
$305.17 = 305$
$2.1902 = 2.19$
$100.9 = 101$
$99.999 = 100$
$55555.5 = 55{,}600$
$672{,}510 = 673{,}000$
$33.333 = 33.3$
$6.66666 = 6.67$

34. Formula: $V = I \times R$
Substitution: $V = 5 \text{ A} \times 20 \ \Omega$
Answer: $V = 100 \text{ V}$

35. Formula: $V = I \times R$
Substitution: $V = 0.5 \text{ A} \times 80 \ \Omega$
Answer: $V = 40 \text{ V}$

36. Formula: $I = \dfrac{V}{R}$
Substitution: $I = \dfrac{9 \text{ V}}{100 \ \Omega}$
Answer: $I = 0.09 \text{ A} = 90 \text{ mA}$

37. Formula: $I = \dfrac{V}{R}$
Substitution: $I = \dfrac{10 \text{ V}}{400 \ \Omega}$
Answer: $I = 0.025 \text{ A} = 25 \text{ mA}$

38. Formula: $R = \dfrac{V}{I}$
Substitution: $R = \dfrac{12 \text{ V}}{0.6 \text{ A}}$
Answer: $R = 20 \ \Omega$

39. Formula: $R = \dfrac{V}{I}$
Substitution: $R = \dfrac{3 \text{ V}}{0.012 \text{ A}}$
Answer: $R = 250 \ \Omega$

40. Formula: $P = I \times V$
Substitution: $P = 0.4 \text{ A} \times 12 \text{ V}$
Answer: $P = 4.8 \text{ W}$

41. Formula: $P = I \times V$
Substitution: $P = 0.05 \text{ A} \times 6 \text{ V}$
Answer: $P = 0.3 \text{ W}$

42. Formula: $I = \dfrac{P}{V}$
Substitution: $I = \dfrac{100 \text{ W}}{50 \text{ V}}$
Answer: $I = 2 \text{ A}$

43. Formula: $I = \dfrac{P}{V}$
Substitution: $I = \dfrac{60 \text{ W}}{120 \text{ V}}$
Answer: $I = 0.5 \text{ A}$

44. Formula: $V = \dfrac{P}{I}$
Substitution: $V = \dfrac{1200 \text{ W}}{0.04 \text{ A}}$
Answer: $V = 30{,}000 \text{ V} = 30 \text{ kV}$

45. Formula: $V = \dfrac{P}{I}$
Substitution: $V = \dfrac{44 \text{ W}}{2 \text{ A}}$
Answer: $V = 22 \text{ V}$

46. Formula: $P = I^2 \times R$
Substitution: $P = (7 \text{ A})^2 \times 10 \ \Omega$
Answer: $P = 490 \text{ W}$

47.   Formula: $P = I^2 \times R$
Substitution: $P = (5 \text{ A})^2 \times 50 \text{ Ω}$
Answer: $P = 1250 \text{ W}$

48.   Formula: $R = \dfrac{P}{I^2}$

Substitution: $R = \dfrac{360 \text{ W}}{(6 \text{ A})^2}$

Answer: $R = 10 \text{ Ω}$

49.   Formula: $R = \dfrac{P}{I^2}$

Substitution: $R = \dfrac{400 \text{ W}}{(2 \text{ A})^2}$

Answer: $R = 100 \text{ Ω}$

50.   Formula: $I = \sqrt{\dfrac{P}{R}}$

Substitution: $I = \sqrt{\dfrac{16 \text{ W}}{4 \text{ Ω}}}$

Answer: $I = 2 \text{ A}$

51.   Formula: $I = \sqrt{\dfrac{P}{R}}$

Substitution: $I = \sqrt{\dfrac{180 \text{ W}}{20 \text{ Ω}}}$

Answer: $I = 3 \text{ A}$

52.   Formula: $P = \dfrac{V^2}{R}$

Substitution: $P = \dfrac{(30 \text{ V})^2}{100 \text{ Ω}}$

Answer: $P = 9 \text{ W}$

53.   Formula: $P = \dfrac{V^2}{R}$

Substitution: $P = \dfrac{(6 \text{ V})^2}{72 \text{ Ω}}$

Answer: $P = 0.5 \text{ W}$

54.   Formula: $R = \dfrac{V^2}{P}$

Substitution: $R = \dfrac{(12 \text{ V})^2}{20 \text{ W}}$

Answer: $R = 7.2 \text{ Ω}$

55.   Formula: $R = \dfrac{V^2}{P}$

Substitution: $R = \dfrac{(5 \text{ V})^2}{0.2 \text{ W}}$

Answer: $R = 125 \text{ Ω}$

56.   Formula: $V = \sqrt{P \times R}$
Substitution: $V = \sqrt{20 \text{ W} \times 20 \text{ Ω}}$
Answer: $V = 20 \text{ V}$

57.   Formula: $V = \sqrt{P \times R}$
Substitution: $V = \sqrt{20 \text{ W} \times 5 \text{ Ω}}$
Answer: $V = 10 \text{ V}$

58.   Formula: cost = kW × h × \$/kWh
Substitution: cost = 8 × 0.04 kW × 3 h ×
            \$0.12/kWh
Answer: cost = \$0.12

59.   Formula: cost = kW × h × \$/kWh
Substitution: cost = 0.12 kW × 2 h × \$0.06/kWh
Answer: cost = \$0.01

60.   Formula: cost = kW × h × \$/kWh
Substitution: cost = 1.4 kW × 0.5 h × \$0.10/kWh
Answer: cost = \$0.07

61. $R_T = 551 \text{ kΩ}$

62. $V_T = 99.85 \text{ mV}$

63.   Formula: $V = I \times R$
Substitution: $V = 25 \text{ mA} \times 4 \text{ kΩ}$
Answer: $V = 100 \text{ V}$

64.   Formula: $I = \dfrac{V}{R}$

Substitution: $I = \dfrac{20 \text{ V}}{5 \text{ kΩ}}$

Answer: $I = 4 \text{ mA}$

65.   Formula: $P = I^2 \times R$
Substitution: $P = (50 \text{ mA})^2 \times 8 \text{ Ω}$
Answer: $P = 0.02 \text{ W} = 20 \text{ mW}$

ANSWERS TO CHAPTER TEST IN THE INSTRUCTOR'S MANUAL

Pages 43–50

1. Voltage: The driving force or potential energy.
   letter symbol: *V*
   unit of measure: volt
   unit symbol: V
2. Current: The flow of electricity.
   letter symbol: *I*
   unit of measure: amp
   unit symbol: A
3. Resistance: The opposition to the flow of electrical current.
   letter symbol: *R*
   unit of measure: ohm
   unit symbol: Ω
4. Power: Electrical work performed.
   letter symbol: *P*
   unit of measure: watt
   unit symbol: W
5. Alternating current: Current in an electric circuit that periodically changes direction due to the voltage changing polarity.
6. Polarity: The positive and negative relationships of a voltage.
7. Resistor: Electrical component used to oppose the flow of electricity.
8. Potential difference: The voltage measured between two points in a circuit.
9. Direct current: Current that always flows in one direction.
10. Electromotive force: Electrical pressure. Another name for voltage.
11. 43,100
    231
    201
    77,800
    33.3

12. Formula: $V = I \times R$
    Substitution: $V = 4 \text{ A} \times 15 \text{ Ω}$
    Answer: $V = 60 \text{ V}$

13. Formula: $I = \dfrac{V}{R}$

    Substitution: $I = \dfrac{9 \text{ V}}{30 \text{ Ω}}$

    Answer: $I = 0.3 \text{ A}$

14. Formula: $R = \dfrac{V}{I}$

    Substitution: $R = \dfrac{20 \text{ V}}{0.5 \text{ A}}$

    Answer: $R = 40 \text{ Ω}$

15. Formula: $P = I \times V$
    Substitution: $P = 0.8 \text{ A} \times 6 \text{ V}$
    Answer: $P = 4.8 \text{ W}$

16. Formula: $I = \dfrac{P}{V}$

    Substitution: $I = \dfrac{60 \text{ W}}{120 \text{ V}}$

    Answer: $I = 0.5 \text{ A}$

17. Formula: $V = \dfrac{P}{I}$

    Substitution: $V = \dfrac{16 \text{ W}}{2 \text{ A}}$

    Answer: $V = 8 \text{ V}$

18. Formula: $P = I^2 \times R$
    Substitution: $P = (4 \text{ A})^2 \times 8 \text{ Ω}$
    Answer: $P = 128 \text{ W}$

19. Formula: $P = \dfrac{V^2}{R}$

    Substitution: $P = \dfrac{(30 \text{ V})^2}{100 \text{ Ω}}$

    Answer: $P = 9 \text{ W}$

20. Formula: $R = \dfrac{V^2}{P}$

    Substitution: $R = \dfrac{(6 \text{ V})^2}{9 \text{ W}}$

    Answer: $R = 4 \text{ Ω}$

21. Formula: cost = kW × hours × price
    Substitution: cost = 6 kW × 4 hours × 0.06
    Answer: cost = $1.44

22. 641.7 kΩ

23. 450.3 mV

24. Formula: $V = I \times R$
     Substitution: $V = 50 \text{ mA} \times 4 \text{ k}\Omega$
     Answer: $V = 200$ V

25. Formula: $I = \dfrac{V}{R}$

     Substitution: $I = \dfrac{50 \text{ V}}{2.5 \text{ k}\Omega}$

     Answer: $I = 20$ mA

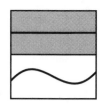

Chapter Test

4

Name: _____

Date: _____

Class: _____

Voltage, Current, Resistance, and Power

For the following terms, supply a definition and the requested information.

1. Voltage: _____

   letter symbol_____   unit of measure_____   unit symbol_____

2. Current: _____

   letter symbol_____   unit of measure_____   unit symbol_____

3. Resistance: _____

   letter symbol_____   unit of measure_____   unit symbol_____

4. Power: _____

   letter symbol_____   unit of measure_____   unit symbol_____

Define the following technical terms.

5. Alternating current: _____

6. Polarity: _____

7. Resistor: _____

8. Potential difference: _____

9. Direct current: _____

10. Electromotive force: _____

11. Round the following numbers to three significant figures.

    43,056 = _____

    231.38 = _____

    200.9 = _____

    77,755.5 = _____

    33.333 = _____

With each problem, write the formula, substitution, and answer.

12. How much voltage is needed to produce four amps of current through a resistive load of 15 ohms?

Formula: _____

Substitution: _____

Answer: _____

13. Calculate the current in a circuit with a nine volt battery connected to a lamp having a resistance of 30 ohms.

Formula: _____

Substitution: _____

Answer: _____

14. What is the resistance in a circuit when 0.5 amps flow from a 20 volt source?

Formula: _____

Substitution: _____

Answer: _____

15. An ammeter connected in a circuit measures 0.8 amps. How much power is produced from the six volt battery?

Formula: _____

Substitution: _____

Answer: _____

16. How much current flows through a 60 watt light bulb when it is connected to a 120 volt supply?

Formula: _____

Substitution: _____

Answer: _____

17. What is the voltage applied to a circuit with two amps of current in a 16 watt load?

Formula: _____

Substitution: _____

Answer: _____

18. Calculate the power consumed in an eight ohm load with a current of four amps.

Formula: _____

Substitution: _____

Answer: _____

19. If a 100 ohm resistor is connected to a 30 volt source, what is the power dissipated?

Formula: _____

Substitution: _____

Answer: _____

20. Determine the resistance of a nine watt light with six volts.

Formula: _____

Substitution: _____

Answer: _____

21. How much will it cost to operate six 1000 watt lights for four hours with a cost of six cents per kWh?

Formula: _____

Substitution: _____

Answer: _____

22. Add these three resistors connected in series to find their combined resistance: 2.7 kilohm, 39 kilohm, 0.6 megohm. Show work here.

Answer: _____

23. If a 450.6 millivolt microphone signal is reduced by 300 microvolts, how much voltage remains? Show work here.

Answer: _____

24. Use Ohm's law to find the voltage in a circuit with 50 milliamps of current through a load of four kilohms.

Formula: _____

Substitution: _____

Answer: _____

25. Calculate the current in a 50 volt circuit with a load of 2.5 kilohms.

Formula: _____

Substitution: _____

Answer: _____

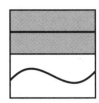

5

Circuit Components

OBJECTIVES

After studying this chapter, students should be able to:
- Identify the characteristics of conductors.
- Describe how to select wire for an electrical conductor.
- Explain the ratings of fuses and circuit breakers.
- Identify the style and ratings of switches.
- Determine the value of resistors using the color code.
- Describe the ratings of variable resistors.

INSTRUCTIONAL MATERIALS

Text: Pages 109–142
Test Your Knowledge Questions, Pages 140–141
Study Guide: Pages 35–40
Laboratory Manual: Pages 41–46

ANSWERS TO TEXTBOOK

Test Your Knowledge, Pages 140–141

1. A conductor offers little or no resistance to an electric current. An insulator offers great or infinite resistance to current.
2. Material, cross-sectional area, resistivity, other variables.
3. Silver, copper, gold, aluminum.
4. A wire with a small cross-sectional area will carry less current than a wire with a large cross-sectional area.
5. R = 12.9 ohms.
6. G = 0.2 S (or mho).
7. As temperature decreases, wire resistance decreases.
8. AWG #12.
9. Solid.
10. Maximum current, reaction time, maximum voltage.
11. Toggle.
12. Push-button.
13. Refer to figures 5-23 to 5-26.
14. Wattage rating and ohmic value.
15. Wattage rating and ohmic value.
16. Note: answers given here are the numbers for the four colors and the nominal value.
    a. 3-3-2-10% = 3300 Ω = 3.3 kΩ ±10%
    b. 1-0-1-5% = 100 Ω ±5%
    c. 2-7-4-10% = 270,000 = 270 kΩ ±10%
    d. 5-9-3-10% = 59,000 = 59 kΩ ±5%
    e. 4-6-0-20% = 46 Ω ±20%
17. a. orange - white - brown
    b. red - violet - red
    c. brown - black - orange
    d. orange - white - black
    e. brown - black - green
18. a. nominal: 1800 Ω ±5%
       range: 90 Ω
       maximum: 1890 Ω
       minimum: 1710 Ω
    b. nominal: 65,000 Ω ±10%
       range: 6500 Ω
       maximum: 71,500 Ω
       minimum: 58,500 Ω
    c. nominal: 90 Ω ±5%
       range: 4.5 Ω
       maximum: 94.5 Ω
       minimum: 85.5 Ω
    d. nominal: 220,000 Ω ±10%
       range: 22,000 Ω
       maximum: 242,000 Ω
       minimum: 198,000 Ω
    e. nominal: 390 ±20%
       range: 78 Ω
       maximum: 468 Ω
       minimum: 312 Ω

ANSWERS TO STUDY GUIDE

Pages 35–40

1. h.
2. c.
3. j.
4. b.
5. l.
6. e.
7. n.
8. o.
9. m.
10. a.
11. g.
12. d.
13. f.
14. i.
15. k.
16. p.
17. a. Material.
    b. Cross-sectional area.
    c. Resistivity/conductivity.
    d. Other variables.
18. a. Silver.
    b. Copper.
    c. Gold.
    d. Aluminum.
19. c.
20. Formula: $R = \rho \dfrac{L}{A}$

    Substitution: $R = 10.37 \times \dfrac{3000 \text{ ft.}}{1620 \text{ cmil}}$

    Answer: $R = 19.2 \ \Omega$

    Alternate Method:
    Formula: $R = $ length $\times$ ohms per 100 feet
    Substitution: $R = 3000$ ft. $\times 0.651 \ \Omega/100$ ft.
    Answer: $R = 19.53 \ \Omega$

21. Formula: $G = \dfrac{1}{R}$

    Substitution: $G = \dfrac{1}{2 \ \Omega}$

    Answer: $G = 0.5$ mho
22. An increase in temperature will increase the resistance.
23. #20 AWG = 3 amps
    #18 AWG = 5 amps
    #14 AWG = 15 amps
    #12 AWG = 20 amps
    #10 AWG = 30 amps

24. b.
25. a. Maximum current.
    b. Reaction time.
    c. Maximum voltage.
26. c.
27. Refer to textbook figures 5-23 to 5-26.
28. wattage and ohmic value
29. wattage and ohmic value
30. Note: answers given here are the numbers for the four colors and the nominal value.
    a. 3-9-1-10% = 390 Ω ± 10%
    b. 2-0-4-5% = 200,000 Ω = 200 kΩ ±5%
    c. 1-2-3-10% = 12,000 Ω = 12 kΩ ±10%
    d. 4-7-2-5% = 4700 Ω = 4.7 kΩ ±5%
    e. 9-1-0.1-5% = 9.1 Ω ±5%
    f. 6-5-2-20% = 6500 Ω = 6.5 kΩ ±20%
    g. 5-3-3-10% = 53,000 Ω = 53 kΩ ±10%
    h. 8-7-0.01-5% = 0.87 Ω ±5%
    i. 7-0-0-5% = 70 Ω ±5%
    j. 1-0-5-10% = 1,000,000 Ω = 1 MΩ ±10%
31. a. brown - black - brown
    b. red - violet - red
    c. orange - orange - orange
    d. green - blue - brown
    e. yellow - white - red
    f. brown - red - green
    g. blue - gray - red
    h. brown - black - orange
    i. orange - white - gold
    j. brown - green - black
32. a. nominal: 39 Ω ±5%
        range: 1.95 Ω
        maximum: 40.95 Ω
        minimum: 37.05 Ω
    b. nominal: 2800 Ω ±10%
        range: 280 Ω
        maximum: 3080 Ω
        minimum: 2520 Ω
    c. nominal: 10,000 Ω ±5%
        range: 500 Ω
        maximum: 10,500 Ω
        minimum: 9500 Ω
    d. nominal: 470 ±20%
        range: 94 Ω
        maximum: 564 Ω
        minimum: 376 Ω
    e. nominal: 33 ±5%
        range: 1.65 Ω
        maximum: 34.65 Ω
        minimum: 31.35 Ω

ANSWERS TO CHAPTER TEST IN THE INSTRUCTOR'S MANUAL

Pages 55–57

1. AWG: Used to measure wire sizes.
2. Mho: Used to measure conductivity.
3. Siemen: Used to measure conductivity.
4. Conductivity: The ease with which a conductor allows electron flow.
5. Nominal value: The ideal value.
6. Silver.
   Copper.
   Gold.
7. b.
8. increase

9. Formula: $G = \dfrac{1}{R}$

   Substitution: $G = \dfrac{1}{5\ \Omega}$

   Answer: $G = 0.2$ S

10. a. 20 A
    b. 15 A
    c. 3 A
11. 930 Ω ±10%
12. 2400 Ω ±5%
13. 650 kΩ ±20%
14. brown - gray - brown
15. orange - white - red
16. yellow - violet - orange
17. brown - black - black
18. brown - red - green
19. nominal = 10 kΩ ±5%
    range of tolerance = 500 Ω
    maximum = 10,500 Ω
    minimum = 9500Ω
20. nominal = 33 kΩ ±10%
    range of tolerance = 3300 Ω
    maximum = 36,300 Ω
    minimum = 29,700 Ω

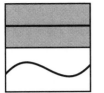

Chapter Test

5

Name: _____

Date: _____

Class: _____

Circuit Components

List the quantities that are measured by the following units.

1. AWG: _____

2. Mho: _____

3. Siemen: _____

Provide the definitions.

4. Conductivity: _____

5. Nominal value: _____

6. List the top three best conductor materials (from lowest to highest conductivity). _____

_____

_____

_____

_____   7. Which wire size has the largest cross-sectional area?
   a. #22 AWG.
   b. #12 AWG.
   c. #14 AWG.

8. As the temperature of a length of wire increases, the resistance of the wire will _____.

9. Calculate the conductance of a length of wire with a resistance of 5 ohms.

Formula: _____

Substitution: _____

Answer: _____

10. What is the current rating for these wire sizes?

    a. #12 AWG _____ amps.

    b. #14 AWG _____ amps.

    c. #20 AWG _____ amps.

Find the nominal values of the resistors.

11. white - orange - brown - silver _____

12. red - yellow - red - gold _____

13. blue - green - yellow - no color _____

Find the first three color bands.

14. 180 ohms   _____ - _____ - _____

15. 3.9 kilohms   _____ - _____ - _____

16. 47 kilohms   _____ - _____ - _____

17. 10 ohms   _____ - _____ - _____

18. 1.2 megohms _____ - _____ - _____

Use the color code to determine the nominal value, range of tolerance, maximum value, and minimum value.

19. brown - black - orange - gold

   nominal = _____   range of tolerance = _____

   maximum = _____   minimum = _____

20. orange - orange - orange - silver

   nominal = _____   range of tolerance = _____

   maximum = _____   minimum = _____

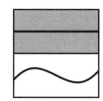
Series Circuits

OBJECTIVES

After studying this chapter, students should be able to:
- Identify a series circuit by its schematic diagram and the circuit current.
- Calculate total resistance in a series circuit when given individual resistance values or voltage and current.
- Calculate the voltage drops in a series circuit.
- Predict the effects of large and small resistors on other parts of the circuit.
- Use the voltage divider formula to determine circuit voltages.
- Calculate the power in a series circuit.
- Recognize the symptoms created by opens and shorts in a series circuit.

INSTRUCTIONAL MATERIALS

Text: Pages 143–180
          Test Your Knowledge Questions, Pages 173–180
Study Guide: Pages 41–51
Laboratory Manual: Pages 47–56

ANSWERS TO TEXTBOOK

Test Your Knowledge, Pages 173–180
1. Student responses will vary.
2. a. Only one path for the current.
    b. Voltage will drop across each resistor.
    c. Total resistance is the sum of the individual resistances.
    d. Total power is the sum of the individual powers.
3. $R_T = R_1 + R_2 + R_3 + ...R_N$
4. Add the known resistance values and subtract from the total.
    Example: $R_2 = R_T - (R_1 + R_3)$

5. b.
6. a.
7. b.
8. The large resistor is so large that almost all of the voltage will drop across it. The other resistors can be ignored.
9. A very small resistor will have so little effect it can usually be ignored.

10. $V_{R_1} = V_A \times \dfrac{R_1}{R_1 + R_2}$

11. Open.
12. One light socket is shorted.
13. There is an open in the fifth bulb.
14. $R_T = 5 \text{ k}\Omega$
15. $R_3 = 600 \ \Omega$

16. $R_1 = 390 \ \Omega$
    $R_2 = 220 \ \Omega$
    $R_3 = 470 \ \Omega$
    a. $R_T = 1080 \ \Omega$
    b. $I = 50 \text{ mA}$

17. a. $R_T = 80 \ \Omega$
    b. $V = 16 \text{ V}$

18. a. $R_T = 30 \text{ k}\Omega$
    b. $R_1 = 5 \text{ k}\Omega$

19. $R_1 = 560 \ \Omega$
    $R_2 = 970 \ \Omega$
    $R_3 = 800 \ \Omega$
    a. $R_T = 2330 \ \Omega$
    b. $I = 19.7 \text{ mA}$
    c. $V_{R_1} = 11 \text{ V}$
       $V_{R_2} = 19.1 \text{ V}$
       $V_{R_3} = 15.8 \text{ V}$

20. a. $I = 0.2$ mA
    b. $R_1 = 5$ kΩ
    c. $R_T = 60$ kΩ
    d. $V = 12$ V

21. a. $R_T = 8$ kΩ
    b. $R_2 = 1$ kΩ
    c. $R_3 = 4$ kΩ

22. $R_1 = 10$ Ω
    $R_2 = 35$ Ω
    $R_3 = 25$ Ω

    a. $R_T = 70$ Ω

    b. $I = 0.3$ A

    c. $P_{R_1} = 0.9$ W
       $P_{R_2} = 3.15$ W
       $P_{R_3} = 2.25$ W

    d. $P_T = 6.3$ W

23. a. $V_1 = V_A \times \dfrac{R_1}{R_1 + R_2}$

       $V_1 = 25$ V

    b. $V_2 = V_A \times \dfrac{R_2}{R_1 + R_2}$

       $V_2 = 75$ V

24. $V_X = V_A \times \dfrac{R_X}{R_T}$

    a. $V_1 = 100$ V
    b. $V_2 = 90$ V
    c. $V_3 = 50$ V
    d. $V_4 = 25$ V

25. a. $R_T = 50$ Ω
    b. $I = 0.8$ A
    c. $V_{R_1} = 8$ V
    d. $P_{R_1} = 6.4$ W
    e. $V_{R_2} = 16$ V
    f. $P_{R_2} = 12.8$ W
    g. $V_{R_3} = 16$ V
    h. $P_{R_3} = 12.8$ W
    i. $P_T = 32$ W

26. a. $I = 0.1$ A
    b. $R_2 = 90$ Ω
    c. $R_1 = 10$ Ω
    d. $P_{R_1} = 0.1$ W
    e. $V_{R_3} = 10$ V
    f. $R_3 = 100$ Ω
    g. $R_T = 200$ Ω
    h. $V = 20$ V
    i. $P_T = 2$ W

ANSWERS TO STUDY GUIDE

Pages 41–51

1. b.
2. c.
3. a.
4. a. Only one path for the current to flow.
   b. Voltage will drop across each resistor.
   c. Total resistance is the sum of the individual resistances.
   d. Total power is the sum of the individual powers.
5. $R_T = R_1 + R_2 + R_3 + \ldots R_N$
6. Add the known resistance values and subtract from the total.
   Example: $R_2 = R_T - (R_1 + R_3 + R_4)$
7. b.
8. a.
9. b.
10. a.
11. b.
12. a.
13. b.
14. a.

15. Formula: $R_T = R_1 + R_2 + R_3$
    Substitution: $R_T = 108$ kΩ + 3.9 kΩ + 1300 Ω
    Answer: $R_T = 113.2$ kΩ

16. Formula: $R_1 = R_T - (R_2 + R_3)$
    Substitution: $R_1 = 150$ kΩ – (35 kΩ + 50 kΩ)
    Answer: $R_1 = 65$ kΩ

17. Resistance values:
    $R_1 = 12$ kΩ
    $R_2 = 27$ kΩ
    $R_3 = 33$ kΩ

a. Total resistance:
Formula: $R_T = R_1 + R_2 + R_3$
Substitution: $R_T = 12 \text{ k}\Omega + 27 \text{ k}\Omega + 33 \text{ k}\Omega$
Answer: $R_T = 72 \text{ k}\Omega$

b. Current:

Formula: $I = \dfrac{V}{R_T}$

Substitution: $I = \dfrac{36 \text{ V}}{72 \text{ k}\Omega}$

Answer: $I = 0.5 \text{ mA}$

18. a. Total resistance:
Formula: $R_T = R_1 + R_2 + R_3 + R_4 + R_5$
Substitution: $R_T = 120 \text{ }\Omega + 80 \text{ }\Omega + 100 \text{ }\Omega + 125 \text{ }\Omega + 75 \text{ }\Omega$
Answer: $R_T = 500 \text{ }\Omega$

b. Applied voltage:
Formula: $V = I \times R_T$
Substitution: $V = 60 \text{ mA} \times 500 \text{ }\Omega$
Answer: $V = 30 \text{ V}$

19. a. Toal resistance:

Formula: $R_T = \dfrac{V}{I}$

Substitution: $R_T = \dfrac{6 \text{ V}}{0.2 \text{ A}}$

Answer: $R_T = 30 \text{ }\Omega$

b. Missing resistor, R_2:
Formula: $R_2 = R_T - (R_1 + R_3)$
Substitution: $R_2 = 30 \text{ }\Omega - (5 \text{ }\Omega + 15 \text{ }\Omega)$
Answer: $R_2 = 10 \text{ }\Omega$

20. Resistance values:
$R_1 = 100 \text{ }\Omega$
$R_2 = 680 \text{ }\Omega$
$R_3 = 320 \text{ }\Omega$

a. Total resistance:
Formula: $R_T = R_1 + R_2 + R_3$
Substitution: $R_T = 100 \text{ }\Omega + 680 \text{ }\Omega + 320 \text{ }\Omega$
Answer: $R_T = 1100 \text{ }\Omega$

b. Current:

Formula: $I = \dfrac{V}{R_T}$

Substitution: $I = \dfrac{12 \text{ V}}{1100 \text{ }\Omega}$

Answer: $I = 10.9 \text{ mA}$

c. Voltage drop across R_1:
Formula: $V = I \times R_1$
Substitution: $V = 10.9 \text{ mA} \times 100 \text{ }\Omega$
Answer: $V_{R_1} = 1.09 \text{ V}$

d. Voltage drop across R_2:
Answer: $V_{R_2} = 7.41 \text{ V}$

e. Voltage drop across R_3:
Answer: $V_{R_3} = 3.49 \text{ V}$

21. a. Current:

Formula: $I = \dfrac{V}{R_4}$

Substitution: $I = \dfrac{6 \text{ V}}{500 \text{ }\Omega}$

Answer: $I = 12 \text{ mA}$

b. The value of R_1:

Formula: $R_1 = \dfrac{V}{I}$

Substitution: $R_1 = \dfrac{3 \text{ V}}{12 \text{ mA}}$

Answer: $R_1 = 250 \text{ }\Omega$

c. Total resistance:
Formula: $R_T = R_1 + R_2 + R_3 + R_4 + R_5$
Substitution: $R_T = 250 \text{ }\Omega + 300 \text{ }\Omega + 1.7 \text{ k}\Omega + 500 \text{ }\Omega + 1.5 \text{ k}\Omega$
Answer: $R_T = 4250 \text{ }\Omega$

d. Applied voltage:
Formula: $V = I \times R_T$
Substitution: $V = 12 \text{ mA} \times 4250 \text{ }\Omega$
Answer: $V = 51 \text{ V}$

22. a. Total resistance:

Formula: $R_T = \dfrac{V}{I}$

Substitution: $R_T = \dfrac{80 \text{ V}}{8 \text{ mA}}$

Answer: $R_T = 10 \text{ k}\Omega$

  b. Value of R_2:

Formula: $R_2 = \dfrac{V_2}{I}$

Substitution: $R_2 = \dfrac{16 \text{ V}}{8 \text{ mA}}$

Answer: $R_2 = 2 \text{ k}\Omega$

  c. Value of R_3:
Formula: $R_3 = R_T - (R_1 + R_2)$
Substitution: $R_3 = 10 \text{ k}\Omega - (4 \text{ k}\Omega + 2 \text{ k}\Omega)$
Answer: $R_3 = 4 \text{ k}\Omega$

23. a. Voltage V_1:

Formula: $V_1 = V_A \times \dfrac{R_1}{R_1 + R_2}$

Substitution: $V_1 = 20 \text{ V} \times \dfrac{5 \ \Omega}{5 \ \Omega + 15 \ \Omega}$

Answer: $V_1 = 5 \text{ V}$

  b. Voltage V_2:

Formula: $V_2 = V_A \times \dfrac{R_2}{R_1 + R_2}$

Substitution: $V_2 = 20 \text{ V} \times \dfrac{15 \ \Omega}{5 \ \Omega + 15 \ \Omega}$

Answer: $V_2 = 15 \text{ V}$

24. Resistance values:
$R_1 = 270 \ \Omega$
$R_2 = 330 \ \Omega$
$R_3 = 400 \ \Omega$

  a. Total resistance:
Formula: $R_T = R_1 + R_2 + R_3$
Substitution: $R_T = 270 \ \Omega + 330 \ \Omega + 400 \ \Omega$
Answer: $R_T = 1000 \ \Omega = 1 \text{ k}\Omega$

  b. Current:

Formula: $I = \dfrac{V}{R_T}$

Substitution: $I = \dfrac{10 \text{ V}}{1 \text{ k}\Omega}$

Answer: $I = 10 \text{ mA}$

  c. Power of R_1:
Formula: $P_{R_1} = I^2 \times R_1$
Substitution: $P_{R_1} = (10 \text{ mA})^2 \times 270 \ \Omega$
Answer: $P_{R_1} = 27 \text{ mW}$

  d. Power of R_2:
Answer: $P_{R_2} = 33 \text{ mW}$

  e. Power of R_3:
Answer: $P_{R_3} = 40 \text{ mW}$

  f. Total power:
Formula: $P_T = V \times I$
Substitution: $P_T = 10 \text{ V} \times 10 \text{ mA}$
Answer: $P_T = 100 \text{ mW}$

25. a. Voltage V_1:

Formula: $V_1 = V_A \times \dfrac{R_1 + R_2 + R_3 + R_4}{R_1 + R_2 + R_3 + R_4}$

Substitution: $V_1 = 80 \text{ V} \times \dfrac{80 \ \Omega}{80 \ \Omega}$

Answer: $V_1 = 80 \text{ V}$

  b. Voltage V_2:

Formula: $V_2 = V_A \times \dfrac{R_2 + R_3 + R_4}{R_1 + R_2 + R_3 + R_4}$

Substitution: $V_2 = 80 \text{ V} \times \dfrac{70 \ \Omega}{80 \ \Omega}$

Answer: $V_2 = 70 \text{ V}$

c.   Voltage V_3:

Formula: $V_3 = V_A \times \dfrac{R_3 + R_4}{R_1 + R_2 + R_3 + R_4}$

Substitution: $V_3 = 80 \text{ V} \times \dfrac{40\ \Omega}{80\ \Omega}$

Answer: $V_3 = 40 \text{ V}$

d.   Voltage V_4:

Formula: $V_4 = V_A \times \dfrac{R_4}{R_1 + R_2 + R_3 + R_4}$

Substitution: $V_4 = 80 \text{ V} \times \dfrac{20\ \Omega}{80\ \Omega}$

Answer: $V_4 = 20 \text{ V}$

e.   Voltage V_5:
Formula: Use applied voltage.
Substitution/Answer: $V_5 = V_B = $ -5 V

f.   Total voltage V_T:
Formula: $V_T = V_A + V_B$
Substitution: $V_T = 80 \text{ V} + 5 \text{ V}$
Answer: $V_T = 85 \text{ V}$

26.   a.   Total resistance:
Formula: $R_T = R_1 + R_2 + R_3$
Substitution: $R_T = 25\ \Omega + 75\ \Omega + 100\ \Omega$
Answer: $R_T = 200\ \Omega$

b.   Current:

Formula: $I = \dfrac{V}{R_T}$

Substitution: $I = \dfrac{100 \text{ V}}{200\ \Omega}$

Answer: $I = 0.5 \text{ A}$

c.   Voltage drop across R_1:
Formula: $V_{R_1} = I \times R_1$
Substitution: $V_{R_1} = 0.5 \text{ A} \times 25\ \Omega$
Answer: $V_{R_1} = 12.5 \text{ V}$

d.   Power of R_1:
Formula: $P_{R_1} = I \times V_{R_1}$
Substitution: $P_{R_1} = 0.5 \text{ A} \times 12.5 \text{ V}$
Answer: $P_{R_1} = 6.25 \text{ W}$

e.   Voltage drop across R_2:
Formula: $V_{R_2} = I \times R_2$
Substitution: $V_{R_2} = 0.5 \text{ A} \times 75\ \Omega$
Answer: $V_{R_2} = 37.5 \text{ V}$

f.   Power of R_2:
Formula: $P_{R_2} = I \times V_{R_2}$
Substitution: $P_{R_2} = 0.5 \text{ A} \times 37.5 \text{ V}$
Answer: $P_{R_2} = 18.75 \text{ W}$

g.   Voltage drop across R_3:
Formula: $V_{R_3} = I \times R_3$
Substitution: $V_{R_3} = 0.5 \text{ A} \times 100\ \Omega$
Answer: $V_{R_3} = 50 \text{ V}$

h.   Power of R_3:
Formula: $P_{R_3} = I \times V_{R_3}$
Substitution: $P_{R_3} = 0.5 \text{ A} \times 50 \text{ V}$
Answer: $P_{R_3} = 25 \text{ W}$

i.   Total power:
Formula: $P_T = V \times I$
Substitution: $P_T = 100 \text{ V} \times 0.5 \text{ A}$
Answer: $P_T = 50 \text{ W}$

27.   a.   Current:

Formula: $I = \dfrac{P_{R_3}}{V_{R_3}}$

Substitution: $I = \dfrac{2.88 \text{ W}}{24 \text{ V}}$

Answer: $I = 0.12 \text{ A}$

b. Value of R_3:

Formula: $R_3 = \dfrac{V_{R_3}}{I}$

Substitution: $R_3 = \dfrac{24 \text{ V}}{0.12 \text{ A}}$

Answer: $R_3 = 200 \text{ }\Omega$

c. Value of R_1:

Formula: $R_1 = \dfrac{P_{R_1}}{I^2}$

Substitution: $R_1 = \dfrac{0.72 \text{ W}}{(0.12 \text{ A})^2}$

Answer: $R_1 = 50 \text{ }\Omega$

d. Voltage drop across R_1:

Formula: $V_{R_1} = I \times R_1$

Substitution: $V_{R_1} = 0.12 \text{ A} \times 50 \text{ }\Omega$

Answer: $V_{R_1} = 6 \text{ V}$

e. Value of R_2:

Formula: $R_2 = \dfrac{V_{R_2}}{I}$

Substitution: $R_2 = \dfrac{30 \text{ V}}{0.12 \text{ A}}$

Answer: $R_2 = 250 \text{ }\Omega$

f. Power dissipated in R_2:

Formula: $P_{R_2} = I \times V_{R_2}$

Substitution: $P_{R_2} = 0.12 \text{ A} \times 30 \text{ V}$

Answer: $P_{R_2} = 3.6 \text{ W}$

g. Total resistance:

Formula: $R_T = R_1 + R_2 + R_3$
Substitution: $R_T = 50 \text{ }\Omega + 250 \text{ }\Omega + 200 \text{ }\Omega$
Answer: $R_T = 500 \text{ }\Omega$

h. Applied voltage:
Formula: $V = I \times R_T$
Substitution: $V = 0.12 \text{ A} \times 500 \text{ }\Omega$
Answer: $V = 60 \text{ V}$

i. Total power:
Formula: $P_T = I \times V$
Substitution: $P_T = 0.12 \text{ A} \times 60 \text{ V}$
Answer: $P_T = 7.2 \text{ W}$

ANSWERS TO CHAPTER TEST IN THE INSTRUCTOR'S MANUAL

Pages 67–77

		Open Circuit	*Short Circuit*
1.	Current.	zero	maximum
2.	Voltage across the defective component.	applied voltage	zero

3. b.
4. a.
5. b.
6. applied voltage
7. small
8. open (or blown)
9. b.
10. a.

11. Formula: $R_T = R_1 + R_2 + R_3$
    Substitution: $R_T = 4.7 \text{ k}\Omega + 25 \text{ k}\Omega + 300 \text{ }\Omega$
    Answer: $R_T = 30 \text{ k}\Omega$

12. Formula: $R_1 = R_T - (R_2 + R_3)$
    Substitution: $R_1 = 270 \text{ k}\Omega - (90 \text{ k}\Omega + 80 \text{ k}\Omega)$
    Answer: $R_1 = 100 \text{ k}\Omega$

13. Color codes: $R_1 = 230 \text{ }\Omega$, $R_2 = 320 \text{ }\Omega$, $R_3 = 140 \text{ }\Omega$

a. Total resistance:
    Formula: $R_T = R_1 + R_2 + R_3$
    Substitution: $R_T = 230 \text{ }\Omega + 320 \text{ }\Omega + 140 \text{ }\Omega$
    Answer: $R_T = 690 \text{ }\Omega$

b. Current:

    Formula: $I = \dfrac{V}{R_T}$

    Substitution: $I = \dfrac{69 \text{ V}}{690 \text{ }\Omega}$

    Answer: $I = 0.1 \text{ A}$

14. a. Total resistance:
    Formula: $R_T = R_1 + R_2 + R_3 + ...R_N$
    Substitution: $R_T = 60\ \Omega + 50\ \Omega + 120\ \Omega +$
    $100\ \Omega + 70\ \Omega$
    Answer: $R_T = 400\ \Omega$

  b. Applied voltage:
    Formula: $V = I \times R_T$
    Substitution: $V = 40\ \text{mA} \times 400\ \Omega$
    Answer: $V = 16\ \text{V}$

15. a. Total resistance:

    Formula: $R_T = \dfrac{V}{I}$

    Substitution: $R_T = \dfrac{12\ \text{V}}{0.3\ \text{A}}$

    Answer: $R_T = 40\ \Omega$

  b. Missing resistor, R_2:
    Formula: $R_2 = R_T - (R_1 + R_3)$
    Substitution: $R_2 = 40\ \Omega - (20\ \Omega + 10\ \Omega)$
    Answer: $R_2 = 10\ \Omega$

16. Color codes: $R_1 = 2000\ \Omega$, $R_2 = 3400\ \Omega$,
    $R_3 = 1300\ \Omega$

  a. Total resistance:
    Formula: $R_T = R_1 + R_2 + R_3$
    Substitution: $R_T = 2\ \text{k}\Omega + 3.4\ \text{k}\Omega + 1.3\ \text{k}\Omega$
    Answer: $R_T = 6.7\ \text{k}\Omega$

  b. Current:

    Formula: $I = \dfrac{V}{R_T}$

    Substitution: $I = \dfrac{67\ \text{V}}{6.7\ \text{k}\Omega}$

    Answer: $I = 10\ \text{mA}$

  c. Voltage of R_1:
    Formula: $V_{R_1} = I \times R_1$
    Substitution: $V_{R_1} = 10\ \text{mA} \times 2\ \text{k}\Omega$
    Answer: $V_{R_1} = 20\ \text{V}$

  d. Voltage of R_2:
    Formula: $V_{R_2} = I \times R_2$
    Substitution: $V_{R_2} = 10\ \text{mA} \times 3.4\ \text{k}\Omega$
    Answer: $V_{R_2} = 34\ \text{V}$

  e. Voltage of R_3:
    Formula: $V_{R_3} = I \times R_3$
    Substitution: $V_{R_3} = 10\ \text{mA} \times 1.3\ \text{k}\Omega$
    Answer: $V_{R_3} = 13\ \text{V}$

17. a. Current:

    Formula: $I = \dfrac{V_{R_4}}{R_4}$

    Substitution: $I = \dfrac{5\ \text{V}}{400\ \Omega}$

    Answer: $I = 12.5\ \text{mA}$

  b. The value of R_1:

    Formula: $R_1 = \dfrac{V_{R_1}}{I}$

    Substitution: $R_1 = \dfrac{10\ \text{V}}{12.5\ \text{mA}}$

    Answer: $R_1 = 800\ \Omega$

  c. Total resistance:
    Formula: $R_T = R_1 + R_2 + R_3 + ...R_N$
    Substitution: $R_T = 800\ \Omega + 600\ \Omega + 1.2\ \text{k}\Omega$
    $+ 400\ \Omega + 1.8\ \text{k}\Omega$
    Answer: $R_T = 4.8\ \text{k}\Omega$

  d. Applied voltage:
    Formula: $V = I \times R_T$
    Substitution: $V = 12.5\ \text{mA} \times 4.8\ \text{k}\Omega$
    Answer: $V = 60\ \text{V}$

18. a. Total resistance:

    Formula: $R_T = \dfrac{V}{I}$

    Substitution: $R_T = \dfrac{12\ \text{V}}{4\ \text{mA}}$

    Answer: $R_T = 3\ \text{k}\Omega$

b. Value of R_2:

Formula: $R_2 = \dfrac{V_{R_2}}{I}$

Substitution: $R_2 = \dfrac{3.2 \text{ V}}{4 \text{ mA}}$

Answer: $R_2 = 800 \text{ } \Omega$

c. Value of R_3:
Formula: $R_3 = R_T - (R_1 + R_2)$
Substitution: $R_3 = 3 \text{ k}\Omega - (1 \text{ k}\Omega + 800 \text{ } \Omega)$
Answer: $R_3 = 1.2 \text{ k}\Omega$

19. a. Voltage V_1:

Formula: $V_1 = V_A \times \dfrac{R_1}{R_1 + R_2}$

Substitution: $V_1 = 40 \text{ V} \times \dfrac{10 \text{ } \Omega}{10 \text{ } \Omega + 30 \text{ } \Omega}$

Answer: $V_1 = 10 \text{ V}$

b. Voltage V_2:

Formula: $V_2 = V_A \times \dfrac{R_2}{R_1 + R_2}$

Substitution: $V_2 = 40 \text{ V} \times \dfrac{30 \text{ } \Omega}{10 \text{ } \Omega + 30 \text{ } \Omega}$

Answer: $V_2 = 30 \text{ V}$

20. Multi-tap formula:

Formula: $V_X = V_A \times \dfrac{R_X}{R_T}$

Total resistance:
Formula: $R_T = R_1 + R_2 + R_3 + R_4$
Substitution: $R_T = 20 \text{ } \Omega + 40 \text{ } \Omega + 10 \text{ } \Omega + 30 \text{ } \Omega$
Answer: $R_T = 100 \text{ } \Omega$
Note that voltages are in reference to ground.

a. Voltage V_1:
Formula/Substitution/Answer: $V_1 = V_A = 20 \text{ V}$

b. Voltage V_2:

Formula: $V_2 = V_A \times \dfrac{R_2 + R_3 + R_4}{R_T}$

Substitution: $V_2 = 20 \text{ V} \times \dfrac{40 \text{ } \Omega + 10 \text{ } \Omega + 30 \text{ } \Omega}{100 \text{ } \Omega}$

Answer: $V_2 = 16 \text{ V}$

c. Voltage V_3:

Formula: $V_3 = V_A \times R_3 + \dfrac{R_4}{R_T}$

Substitution: $V_3 = 20 \text{ V} \times \dfrac{10 \text{ } \Omega + 30 \text{ } \Omega}{100 \text{ } \Omega}$

Answer: $V_3 = 8 \text{ V}$

d. Voltage V_4:

Formula: $V_4 = V_A \times \dfrac{R_4}{R_T}$

Substitution: $V_4 = 20 \text{ V} \times \dfrac{30 \text{ } \Omega}{100 \text{ } \Omega}$

Answer: $V_4 = 6 \text{ V}$

e. Voltage V_5:
Formula/Substitution/Answer: $V_5 = V_B = -25 \text{ V}$

f. Voltage V_T:
Formula: $V_T = V_A + V_B$
Substitution: $V_T = 20 \text{ V} + 25 \text{ V}$
Answer: $V_T = 45 \text{ V}$

Chapter Test

6

Name: _____

Date: _____

Class: _____

Series Circuits

For questions 1 and 2, fill in the information requested on the left for an open circuit and a short circuit.

	Open Circuit	Short Circuit
1. Current.	_____	_____
2. Voltage across the defective component.	_____	_____

Select the best answer.

_____ 3. Current measured at different points in a series circuit:
   a. depends on the value of the resistor.
   b. is the same at all points.
   c. drops as it goes through a resistor.
   d. is highest close to the negative side of the voltage source and lower further away.

_____ 4. Voltage in a series circuit:
   a. drops across each resistor.
   b. is the same at all points.
   c. flows from positive to negative.
   d. None of the above.

_____ 5. Power in a series circuit:
   a. is the same at all points.
   b. is the sum of the individual powers.
   c. flows from positive to negative.
   d. None of the above.

Fill in the blanks.

6. If one resistor is 10 times larger than any other in a series circuit, this resistor will have a voltage drop approximately

   equal to the _____.

7. If one resistor has a very small effect on the amount of current in a series circuit, it is because the ohmic value of the

   resistor is very _____ in comparison to the others.

8. In a series circuit, if all of the applied voltage is dropped across the fuse, it is _____.

Select the best answer.

_____   9. Several light bulbs are connected in series to a 100 volt supply. If one light goes out and the others get
              bright, the problem could be:
              a. a burned-out bulb.
              b. that wires connecting the light are touching.
              c. a wire has broken.
              d. the fuse burned out.

_____ 10. A circuit with five light bulbs is connected in series to a 100 volt supply and all of the lights go out. The
              ammeter reads 0 amps. A voltmeter reads 100 volts at the supply, 0 volts across four of the lights, and
              100 volts across the fifth. What is the possible problem?
              a. A burned-out bulb.
              b. Wires connecting the light are touching.
              c. A wire has broken.
              d. The fuse burned out.

Name: _____

With each problem, write the formula, substitution, and answer.

11. Using the values given in the following circuit, calculate total resistance.

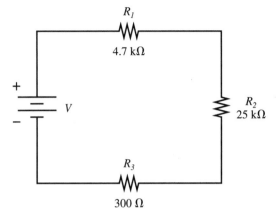

Formula: _____

Substitution: _____

Answer: _____

12. In the following figure, the values of total resistance and two of the individual resistors are given. Find the value of the missing resistor.

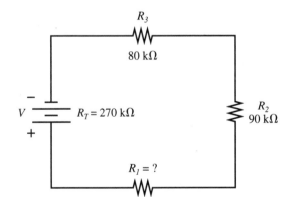

Formula: _____

Substitution: _____

Answer: _____

13. Use the resistor color codes given in the following figure to find the total resistance and current in the circuit.

  a. Total resistance:

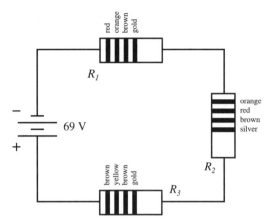

  Formula: _____

  Substitution: _____

  Answer: _____

  b. Current:

  Formula: _____

  Substitution: _____

  Answer: _____

14. Find the total resistance and the applied voltage for the following circuit.

    a. Total resistance:

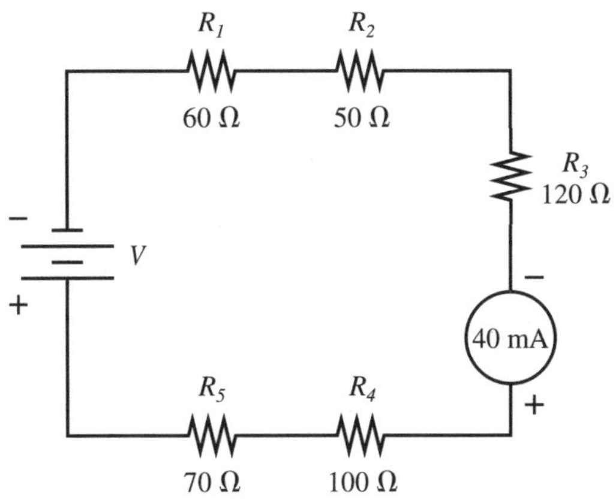

        Formula: _____

        Substitution: _____

        Answer: _____

    b. Applied voltage:

        Formula: _____

        Substitution: _____

        Answer: _____

15. Use the given values of applied voltage and current of the following circuit to find the total resistance and the value of the missing resistor.

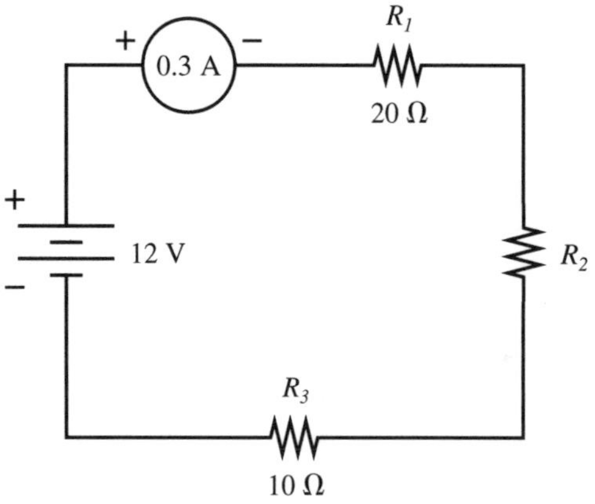

   a. Total resistance:

      Formula: _____

      Substitution: _____

      Answer: _____

   b. Missing resistor, R_2:

      Formula: _____

      Substitution: _____

      Answer: _____

16. Use the resistor color codes given in the following figure to find the total resistance, current, and the individual voltage drops.

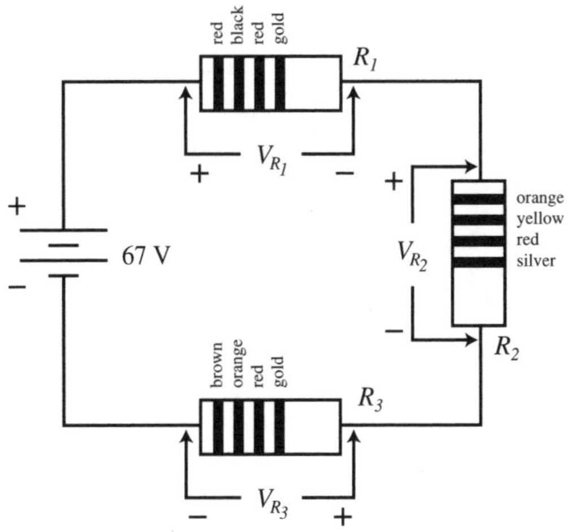

a. Total resistance:

Formula: _____

Substitution: _____

Answer: _____

b. Current:

Formula: _____

Substitution: _____

Answer: _____

c. Voltage of R_1:

Formula: _____

Substitution: _____

Answer: _____

d. Voltage of R_2:

Formula: _____

Substitution: _____

Answer: _____

e. Voltage of R_3:

Formula: _____

Substitution: _____

Answer: _____

Circuit figure:

R_1 (red black red gold), V_{R_1}, 67 V, R_2 (orange yellow red silver), V_{R_2}, R_3 (brown orange red gold), V_{R_3}

17. Use the voltage drops and resistance values given in the following figure to find: current, the value of R_1, total resistance, and the applied voltage.

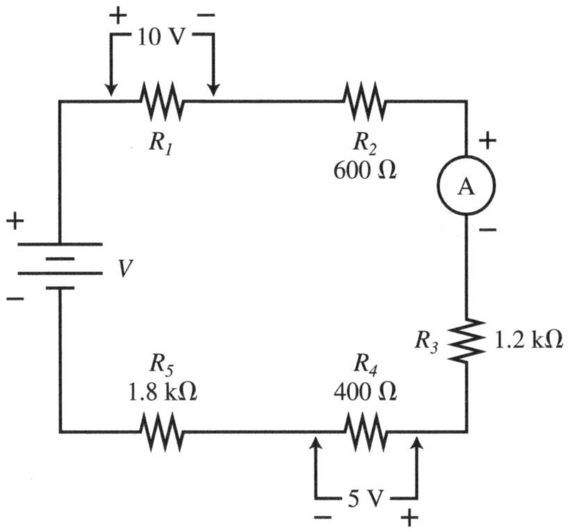

a. Current:

Formula: _____

Substitution: _____

Answer: _____

b. The value of R_1:

Formula: _____

Substitution: _____

Answer: _____

c. Total resistance:

Formula: _____

Substitution: _____

Answer: _____

d. Applied voltage:

Formula: _____

Substitution: _____

Answer: _____

18. Using the values shown in the following figure, find: total resistance, the value of resistor R_2, and the value of resistor R_3.

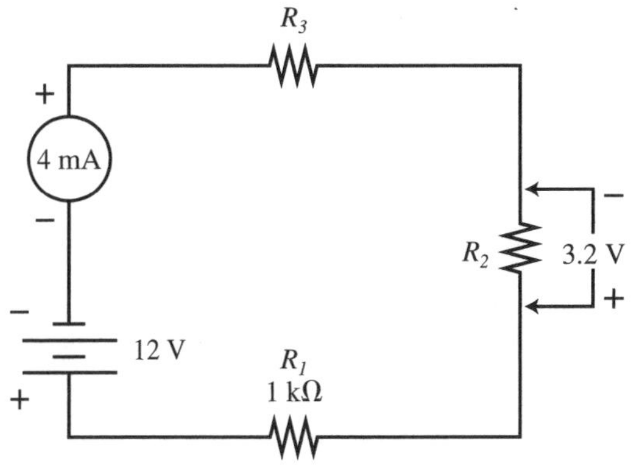

a. Total resistance:

Formula: _____

Substitution: _____

Answer: _____

b. Value of R_2:

Formula: _____

Substitution: _____

Answer: _____

c. Value of R_3:

Formula: _____

Substitution: _____

Answer: _____

19. Use the two voltage divider formulas to find the voltages across the resistors in the following figure.

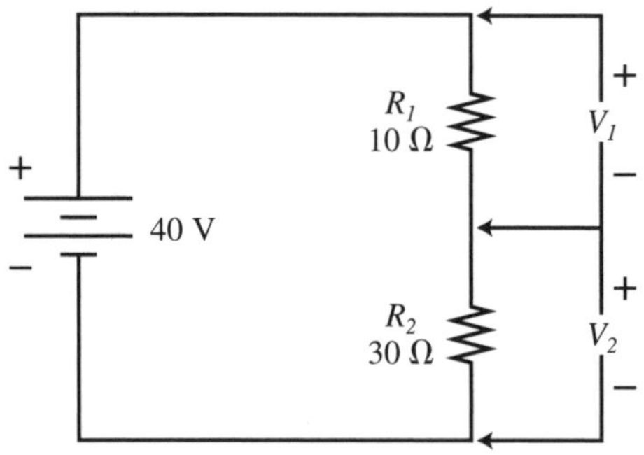

   a. Voltage V_1:

      Formula: _____

      Substitution: _____

      Answer: _____

   b. Voltage V_2:

      Formula: _____

      Substitution: _____

      Answer: _____

20. Use the multi-tap resistor voltage divider formula to find the voltages labeled in the following figure.

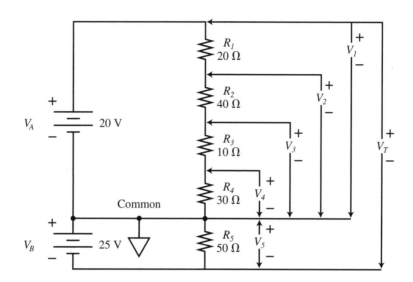

a. Voltage V_1:

Formula: _____

Substitution: _____

Answer: _____

b. Voltage V_2:

Formula: _____

Substitution: _____

Answer: _____

c. Voltage V_3:

Formula: _____

Substitution: _____

Answer: _____

d. Voltage V_4:

Formula: _____

Substitution: _____

Answer: _____

e. Voltage V_5:

Formula: _____

Substitution: _____

Answer: _____

f. Voltage V_T:

Formula: _____

Substitution: _____

Answer: _____

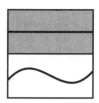

Parallel Circuits

OBJECTIVES

After studying this chapter, students should be able to:
- Recognize and identify loads connected in parallel.
- Calculate the total resistance of a parallel circuit using different formulas.
- Calculate power and current in branch resistances and the total circuit.
- Analyze branch currents to determine mainline currents at various points in a parallel circuit.
- Evaluate symptoms of faulty circuits to determine the problem.

INSTRUCTIONAL MATERIALS

Text: Pages 181–210
      Test Your Knowledge Questions, Pages 206–210
Study Guide: Pages 53–63
Laboratory Manual: Pages 57–70

ANSWERS TO TEXTBOOK

Test Your Knowledge, Pages 206–210
1. Student responses will vary.
2. a. Voltage is the same throughout the circuit.
   b. Current splits to each individual branch.
   c. Total current is the sum of the branch currents.
   d. Total resistance is smaller than the smallest branch resistance.
   e. Total power is the sum of the individual powers.
3. $\dfrac{1}{R_T} = \dfrac{1}{R_1} + \dfrac{1}{R_2} + \dfrac{1}{R_3} + \ldots \dfrac{1}{R_N}$

   $R_T = \dfrac{1}{G_T}$ plus $G_T = G_1 + G_2 + G_3 + \ldots G_N$

$$R_T = \frac{R_1 \times R_2}{R_1 + R_2}$$

$$R_T = \frac{R}{N}$$

4. Total resistance is smaller than the smallest branch resistor.
5. The siemen or the mho.
6. When there are only two resistors in parallel.
7. When there are any number of *equal value* resistors connected in parallel.
8. Total current is the sum of the individual branch currents.
9. Mainline current is the sum of the branch currents from that point away from the voltage source.
10. Total resistance will measure zero.
11. None.

12. a. Reciprocal formula:

   Formula: $\dfrac{1}{R_T} = \dfrac{1}{R_1} + \dfrac{1}{R_2}$

   Substitution: $\dfrac{1}{R_T} = \dfrac{1}{600\ \Omega} + \dfrac{1}{500\ \Omega}$

   Answer: $R_T = 273\ \Omega$

   b. Shortcut formula:

   Formula: $R_T = \dfrac{R_1 \times R_2}{R_1 + R_2}$

   Substitution: $R_T = \dfrac{600\ \Omega \times 500\ \Omega}{600\ \Omega + 500\ \Omega}$

   Answer: $R_T = 273\ \Omega$

13. a.   Total resistance:

Formula: $\dfrac{1}{R_T} = \dfrac{1}{R_1} + \dfrac{1}{R_2} + \dfrac{1}{R_3}$

Substitution: $\dfrac{1}{R_T} = \dfrac{1}{50\ \Omega} + \dfrac{1}{100\ \Omega} + \dfrac{1}{80\ \Omega}$

Answer: $R_T = 23.5\ \Omega$

b.   Total current:

Formula: $I_T = \dfrac{V}{R_T}$

Substitution: $I_T = \dfrac{10\ \text{V}}{23.5\ \Omega}$

Answer: $I_T = 0.426\ \text{A}$

14. a.   Reciprocal formula:

Formula: $\dfrac{1}{R_T} = \dfrac{1}{R_1} + \dfrac{1}{R_2} + \dfrac{1}{R_3}$

Substitution:

$$\dfrac{1}{R_T} = \dfrac{1}{1.2\ \text{k}\Omega} + \dfrac{1}{1.2\ \text{k}\Omega} + \dfrac{1}{1.2\ \text{k}\Omega}$$

Answer: $R_T = 400\ \Omega$

b.   Equal resistors formula:

Formula: $R_T = \dfrac{R}{N}$

Substitution: $R_T = \dfrac{1.2\ \text{k}\Omega}{3}$

Answer: $R_T = 400\ \Omega$

15. a.   Total resistance:

Formula: $\dfrac{1}{R_T} = \dfrac{1}{R_1} + \dfrac{1}{R_2} + \dfrac{1}{R_3} + \dfrac{1}{R_4}$

Substitution: $\dfrac{1}{R_T} = \dfrac{1}{50\ \Omega} + \dfrac{1}{100\ \Omega} +$

$\dfrac{1}{200\ \Omega} + \dfrac{1}{200\ \Omega}$

Answer: $R_T = 25\ \Omega$

b.   Applied voltage:
Formula: $V = I \times R_T$
Substitution: $V = 0.5\ \text{A} \times 25\ \Omega$
Answer: $V = 12.5\ \text{V}$

c.   Total power:
Formula: $P_T = I \times V$
Substitution: $P_T = 0.5\ \text{A} \times 12.5\ \text{V}$
Answer: $P_T = 6.25\ \text{W}$

16. a.   Total resistance:

Formula: $R_T = \dfrac{V}{I}$

Substitution: $R_T = \dfrac{100\ \text{V}}{0.85\ \text{A}}$

Answer: $R_T = 118\ \Omega$

b.   Total power:
Formula: $P_T = I \times V$
Substitution: $P_T = 0.85\ \text{A} \times 100\ \text{V}$
Answer: $P_T = 85\ \text{W}$

c.   Value of R_2:

Formula: $\dfrac{1}{R_2} = \dfrac{1}{R_T} - \left(\dfrac{1}{R_1} + \dfrac{1}{R_3} \right)$

Substitution:

$$\dfrac{1}{R_2} = \dfrac{1}{118\ \Omega} - \left(\dfrac{1}{400\ \Omega} + \dfrac{1}{500\ \Omega} \right)$$

Answer: $R_2 = 250\ \Omega$ (Note: rounding in step a may cause answer to be as high as 252 Ω.)

17. a.   Total resistance:

Formula: $\dfrac{1}{R_T} = \dfrac{1}{R_1} + \dfrac{1}{R_2} + \dfrac{1}{R_3}$

Substitution:

$$\dfrac{1}{R_T} = \dfrac{1}{75\ \text{k}\Omega} + \dfrac{1}{80\ \text{k}\Omega} + \dfrac{1}{60\ \text{k}\Omega}$$

Answer: $R_T = 23.5\ \text{k}\Omega$

b. Total current:

Formula: $I_T = \sqrt{\dfrac{P_T}{R_T}}$

Substitution: $I_T = \sqrt{\dfrac{1.53 \text{ mW}}{23.5 \text{ k}\Omega}}$

Answer: $I_T = 0.255$ mA

c. Applied voltage:

Formula: $V = \dfrac{P_T}{I_T}$

Substitution: $V = \dfrac{1.53 \text{ mW}}{0.255 \text{ mA}}$

Answer: $V = 6$ V

18. a. Applied voltage:
    Formula: $V = I_2 \times R_2$
    Substitution: $V = 30 \text{ mA} \times 2 \text{ k}\Omega$
    Answer: $V = 60$ V

b. Total current:

Formula: $I_T = \dfrac{P_T}{V}$

Substitution: $I_T = \dfrac{7.2 \text{ W}}{60 \text{ V}}$

Answer: $I_T = 0.12$ A

c. Total resistance:

Formula: $R_T = \dfrac{V}{I_T}$

Substitution: $R_T = \dfrac{60 \text{ V}}{0.12 \text{ A}}$

Answer: $R_T = 500$ Ω

d. Branch current I_1:

Formula: $I_1 = \dfrac{V}{R_1}$

Substitution: $I_1 = \dfrac{60 \text{ V}}{2 \text{ k}\Omega}$

Answer: $I_1 = 30$ mA

e. Branch current I_3:
    Formula: $I_3 = I_T - (I_1 + I_2)$
    Substitution: $I_3 = 0.12 \text{ A} - (30 \text{ mA} + 30 \text{ mA})$
    Answer: $I_3 = 60$ mA

19. a. Applied voltage:
    Formula: $V = \sqrt{P_{R_2} \times R_2}$
    Substitution: $V = \sqrt{2.22 \text{ W} \times 72 \text{ }\Omega}$
    Answer: $V = 12.6$ V

b. Total resistance:

Formula: $R_T = \dfrac{V}{I_T}$

Substitution: $R_T = \dfrac{12.6 \text{ V}}{2.5 \text{ A}}$

Answer: $R_T = 5$ Ω

c. Resistor R_3:

Formula: $\dfrac{1}{R_3} = \dfrac{1}{R_T} - \left(\dfrac{1}{R_1} + \dfrac{1}{R_2}\right)$

Substitution: $\dfrac{1}{R_3} = \dfrac{1}{5 \text{ }\Omega} - \left(\dfrac{1}{36 \text{ }\Omega} + \dfrac{1}{72 \text{ }\Omega}\right)$

Answer: $R_3 = 6.3$ Ω

d. Branch current I_1:

Formula: $I_1 = \dfrac{V}{R_1}$

Substitution: $I_1 = \dfrac{12.6 \text{ V}}{36 \text{ }\Omega}$

Answer: $I_1 = 0.35$ A

Branch current I_2:

$I_2 = 0.175$ A

Branch current I_3:

$I_3 = 2$ A

20. Note: to solve this problem, first find the branch currents. Be careful to note the meter number for each branch.

    a. Branch current for R_1 (meter I_3):

        Formula: $I_3 = \dfrac{V}{R_1}$

        Substitution: $I_3 = \dfrac{50 \text{ V}}{25 \text{ }\Omega}$

        Answer: $I_3 = 2$ A

    b. Branch current for R_2 (meter I_6):
        Answer: $I_6 = 0.333$ A

    c. Branch current for R_3 (meter I_8):
        $I_8 = 1$ A
        Note: the remaining currents are found by starting from the branch furthest from the supply and adding as each branch current combines on the mainline buss.

    d. $I_7 = I_8 = 1$ A

    e. $I_4 = I_6 + I_7$
        $I_4 = 1.33$ A

    f. $I_5 = I_4 = 1.33$ A

    g. $I_1 = I_5 + I_4$
        $I_1 = 3.333$ A

    h. $I_2 = I_1 = 3.333$ A

ANSWERS TO STUDY GUIDE

Pages 53–63
1. d.
2. b.
3. e.
4. a.
5. c.
6. a. Voltage is the same throughout the circuit.
    b. Current splits to each individual branch.
    c. Total current is the sum of the branch currents.
    d. Total resistance is smaller than the smallest branch resistance.
    e. Total power is the sum of the individual powers.

7. $\dfrac{1}{R_T} = \dfrac{1}{R_1} + \dfrac{1}{R_2} + \dfrac{1}{R_3} + \dots \dfrac{1}{R_N}$

    $R_T = \dfrac{1}{G_T}$ plus $G_T = G_1 + G_2 + G_3 + \dots G_N$

    $R_T = \dfrac{R_1 \times R_2}{R_1 + R_2}$

    $R_T = \dfrac{R}{N}$

8. b.
9. b.
10. d.
11. a.
12. c.
13. c.
14. a.
15. d.

16. a. Reciprocal formula:

        Formula: $\dfrac{1}{R_T} = \dfrac{1}{R_1} + \dfrac{1}{R_2}$

        Substitution: $\dfrac{1}{R_T} = \dfrac{1}{1.8 \text{ k}\Omega} + \dfrac{1}{2.7 \text{ k}\Omega}$

        Answer: $R_T = 1.08$ kΩ

    b. Shortcut formula:

        Formula: $R_T = \dfrac{R_1 \times R_2}{R_1 + R_2}$

        Substitution: $R_T = \dfrac{1.8 \text{ k}\Omega \times 2.7 \text{ k}\Omega}{1.8 \text{ k}\Omega + 2.7 \text{ k}\Omega}$

        Answer: $R_T = 1.08$ kΩ

17. a. Total resistance:

        Formula: $\dfrac{1}{R_T} = \dfrac{1}{R_1} + \dfrac{1}{R_2} + \dfrac{1}{R_3}$

        Substitution:
        $\dfrac{1}{R_T} = \dfrac{1}{800 \text{ }\Omega} + \dfrac{1}{400 \text{ }\Omega} + \dfrac{1}{800 \text{ }\Omega}$

        Answer: $R_T = 200$ Ω

b. Total current:

Formula: $I_T = \dfrac{V}{R_T}$

Substitution: $I_T = \dfrac{50 \text{ V}}{200 \text{ }\Omega}$

Answer: $I_T = 0.25 \text{ A}$

18. a. Reciprocal formula:

Formula: $\dfrac{1}{R_T} = \dfrac{1}{R_1} + \dfrac{1}{R_2} + \dfrac{1}{R_3}$

Substitution:

$\dfrac{1}{R_T} = \dfrac{1}{18 \text{ M}\Omega} + \dfrac{1}{18 \text{ M}\Omega} + \dfrac{1}{18 \text{ M}\Omega}$

Answer: $R_T = 6 \text{ M}\Omega$

b. Equal resistors formula:

Formula: $R_T = \dfrac{R}{N}$

Substitution: $R_T = \dfrac{18 \text{ M}\Omega}{3}$

Answer: $R_T = 6 \text{ M}\Omega$

c. Total current:

Formula: $I_T = \dfrac{V}{R_T}$

Substitution: $I_T = \dfrac{25 \text{ V}}{6 \text{ M}\Omega}$

Answer: $I_T = 4.17 \text{ }\mu\text{A}$

19. a. Total resistance:

Formula: $\dfrac{1}{R_T} = \dfrac{1}{R_1} + \dfrac{1}{R_2} + \dfrac{1}{R_3} + \dfrac{1}{R_4}$

Substitution:

$\dfrac{1}{R_T} = \dfrac{1}{5 \text{ k}\Omega} + \dfrac{1}{12 \text{ k}\Omega} + \dfrac{1}{7 \text{ k}\Omega} + \dfrac{1}{7 \text{ k}\Omega}$

Answer: $R_T = 1.76 \text{ k}\Omega$

b. Applied voltage:
Formula: $V = I \times R$
Substitution: $V = 11.4 \text{ mA} \times 1.76 \text{ k}\Omega$
Answer: $V = 20 \text{ V}$

c. Total power:
Formula: $P_T = I \times V$
Substitution: $P_T = 11.4 \text{ mA} \times 20 \text{ V}$
Answer: $P_T = 228 \text{ mW}$

20. a. Total resistance:

Formula: $R_T = \dfrac{V}{I}$

Substitution: $R_T = \dfrac{6 \text{ V}}{600 \text{ mA}}$

Answer: $R_T = 10 \text{ }\Omega$

b. Total power:
Formula: $P_T = I \times V$
Substitution: $P_T = 600 \text{ mA} \times 6 \text{ V}$
Answer: $P_T = 3.6 \text{ W}$

c. Value of R_2:

Formula: $\dfrac{1}{R_2} = \dfrac{1}{R_T} - \left(\dfrac{1}{R_1} + \dfrac{1}{R_3} \right)$

Substitution: $\dfrac{1}{R_2} = \dfrac{1}{10 \text{ }\Omega} - \left(\dfrac{1}{47 \text{ }\Omega} + \dfrac{1}{35 \text{ }\Omega} \right)$

Answer: $R_2 = 20 \text{ }\Omega$

21. a. Total resistance:

Formula: $\dfrac{1}{R_T} = \dfrac{1}{R_1} + \dfrac{1}{R_2} + \dfrac{1}{R_3}$

Substitution: $\dfrac{1}{R_T} = \dfrac{1}{24 \text{ }\Omega} + \dfrac{1}{70 \text{ }\Omega} + \dfrac{1}{48 \text{ }\Omega}$

Answer: $R_T = 13 \text{ }\Omega$

b. Formula: $I_T = \sqrt{\dfrac{P_T}{R_T}}$

Substitution: $I_T = \sqrt{\dfrac{52 \text{ W}}{13 \text{ }\Omega}}$

Answer: $I_T = 2 \text{ A}$

c. Applied voltage:

Formula: $V = \dfrac{P_T}{I_T}$

Substitution: $V = \dfrac{52 \text{ W}}{2 \text{ A}}$

Answer: $V = 26$ V

22. a. Applied voltage:
Formula: $V = I_2 \times R_2$
Substitution: $V = 60 \text{ mA} \times 150 \text{ }\Omega$
Answer: $V = 9$ V

b. Total current:

Formula: $I_T = \dfrac{P_T}{V}$

Substitution: $I_T = \dfrac{1.26 \text{ W}}{9 \text{ V}}$

Answer: $I_T = 0.14$ A

c. Total resistance:

Formula: $R_T = \dfrac{V}{I_T}$

Substitution: $R_T = \dfrac{9 \text{ V}}{0.14 \text{ A}}$

Answer: $R_T = 64.3$ Ω

d. Branch current I_1:

Formula: $I_1 = \dfrac{V}{R_1}$

Substitution: $I_1 = \dfrac{9 \text{ V}}{180 \text{ }\Omega}$

Answer: $I_1 = 50$ mA

e. Branch current I_3:
Formula: $I_3 = I_T - (I_1 + I_2)$
Substitution: $I_3 = 140 \text{ mA} - (50 \text{ mA} + 60 \text{ mA})$
Answer: $I_3 = 30$ mA

f. Value of R_3:

Formula: $R_3 = \dfrac{V}{I_3}$

Substitution: $R_3 = \dfrac{9 \text{ V}}{30 \text{ mA}}$

Answer: $R_3 = 300$ Ω

23. a. Applied voltage:
Formula: $V = \sqrt{P_{R_2} \times R_2}$
Substitution: $V = \sqrt{3.6 \text{ W} \times 4000 \text{ }\Omega}$
Answer: $V = 120$ V

b. Total resistance:

Formula: $R_T = \dfrac{V}{I_T}$

Substitution: $R_T = \dfrac{120 \text{ V}}{100 \text{ mA}}$

Answer: $R_T = 1200$ Ω

c. Resistor R_1:

Formula: $\dfrac{1}{R_1} = \dfrac{1}{R_T} - \left(\dfrac{1}{R_2} + \dfrac{1}{R_3} \right)$

Substitution:

$$\dfrac{1}{R_1} = \dfrac{1}{1200 \text{ }\Omega} - \left(\dfrac{1}{4000 \text{ }\Omega} + \dfrac{1}{5000 \text{ }\Omega} \right)$$

Answer: $R_1 = 2609$ Ω

d. Branch current I_1:

Formula: $I_1 = \dfrac{V}{R_1}$

Substitution: $I_1 = \dfrac{120 \text{ V}}{2609 \text{ }\Omega}$

Answer: $I_1 = 46$ mA

e. Branch current I_2:
$I_2 = 30$ mA

f. Branch current I_3:
$I_3 = 24$ mA

24. Note: to solve this problem, first find the branch currents. Be careful to note the meter number for each branch.

Branch current for R_1 (meter I_3):

Formula: $I_3 = \dfrac{V}{R_1}$

Substitution: $I_3 = \dfrac{25 \text{ V}}{50 \text{ }\Omega}$

Answer: $I_3 = 0.5$ A

Branch current for R_2 (meter I_6):
Answer: $I_6 = 1$ A

Branch current for R_3 (meter I_8):
Answer: $I_8 = 2.5$ A

The remaining currents are found by starting from the branch furthest from the supply and adding as each branch current combines on the mainline buss.

Answer: $I_7 = I_8 = 2.5$ A
$I_4 = I_6 + I_7$
Answer: $I_4 = 3.5$ A
Answer: $I_5 = I_4 = 3.5$ A
$I_1 = I_3 + I_4$
Answer: $I_1 = 4$ A
Answer: $I_2 = I_1 = 4$ A

ANSWERS TO CHAPTER TEST IN THE INSTRUCTOR'S MANUAL

Pages 87–95

1. a. Current in the individual branches. They add to equal the total current.
    b. Supply of current closest to the power supply, which carries all of the circuit current.
    c. Voltage is the same across all parallel branches.

2. (Any three of the following:)

    $\dfrac{1}{R_T} = \dfrac{1}{R_1} + \dfrac{1}{R_2} + \dfrac{1}{R_3} + \dots \dfrac{1}{R_N}$

    $R_T = \dfrac{1}{G_T}$

    $R_T = \dfrac{R_1 \times R_2}{R_1 + R_2}$

    $R_T = \dfrac{R}{N}$

3. b.
4. c.
5. d.
6. a.
7. a.
8. d.

9. Formula: $G_T = \dfrac{1}{R_T} = \dfrac{1}{R_1} + \dfrac{1}{R_2}$

    Substitution: $\dfrac{1}{R_T} = \dfrac{1}{3.3 \text{ k}\Omega} + \dfrac{1}{5.4 \text{ k}\Omega}$

    Answer: $R_T = 2.05$ kΩ

10. Formula: $R_T = \dfrac{R_1 \times R_2}{R_1 + R_2}$

    Substitution: $R_T = \dfrac{3.3 \text{ k}\Omega \times 5.4 \text{ k}\Omega}{3.3 \text{ k}\Omega + 5.4 \text{ k}\Omega}$

    Answer: $R_T = 2.05$ kΩ

11. Formula: $\dfrac{1}{R_T} = \dfrac{1}{R_1} + \dfrac{1}{R_2} + \dfrac{1}{R_3}$

    Substitution: $\dfrac{1}{R_T} = \dfrac{1}{60 \text{ }\Omega} + \dfrac{1}{120 \text{ }\Omega} + \dfrac{1}{40 \text{ }\Omega}$

    Answer: $R_T = 20$ Ω

12. Formula: $I = \dfrac{V}{R_T}$

    Substitution: $I = \dfrac{100 \text{ V}}{20 \text{ }\Omega}$

    Answer: $I = 5$ A

13. Formula: $G_T = \dfrac{1}{R_T} = \dfrac{1}{R_1} + \dfrac{1}{R_2} + \dfrac{1}{R_3}$

Substitution:

$$\dfrac{1}{R_T} = \dfrac{1}{12 \text{ M}\Omega} + \dfrac{1}{12 \text{ M}\Omega} + \dfrac{1}{12 \text{ M}\Omega}$$

Answer: $R_T = 4 \text{ M}\Omega$

14. Formula: $R_T = \dfrac{R}{N}$

Substitution: $R_T = \dfrac{12 \text{ M}\Omega}{3}$

Answer: $R_T = 4 \text{ M}\Omega$

15. Formula: $V = \sqrt{P_{R_2} \times R_2}$
Substitution: $V = \sqrt{0.72 \text{ W} \times 200 \text{ }\Omega}$
Answer: $V = 12$ V

16. Formula: $R_T = \dfrac{V}{I_T}$

Substitution: $R_T = \dfrac{12 \text{ V}}{200 \text{ mA}}$

Answer: $R_T = 60 \text{ }\Omega$

17. Formula: $\dfrac{1}{R_T} = \dfrac{1}{R_1} + \dfrac{1}{R_2} + \dfrac{1}{R_3} + \dfrac{1}{R_4}$

Substitution:

$$\dfrac{1}{R_T} = \dfrac{1}{8 \text{ k}\Omega} + \dfrac{1}{10 \text{ k}\Omega} + \dfrac{1}{6 \text{ k}\Omega} + \dfrac{1}{6 \text{ k}\Omega}$$

Answers: $R_T = 1.79 \text{ k}\Omega$

18. Formula: $V = I_T \times R_T$
Substitution: $V = 20.1 \text{ mA} \times 1.79 \text{ k}\Omega$
Answer: $V = 36$ V

19. Formula: $P_T = I_T \times V$
Substitution: $P_T = 20.1 \text{ mA} \times 36 \text{ V}$
Answer: $P_T = 0.724$ W

20. Branch currents: I_3, I_6, I_7, I_8

Formula: $I_3 = \dfrac{V}{R_1}$

Substitution: $I_3 = \dfrac{30 \text{ V}}{60 \text{ }\Omega}$

Answer: $I_3 = 0.5$ A

Answer: $I_6 = 1$ A

Answer: $I_7 = 2$ A

Answer: $I_8 = I_7 = 2$ A

Mainline currents (sum of branch currents past meters I_1 and I_2 = total current):

Formula: $I_T = I_3 + I_6 + I_7$
Substitution: $I_T = 0.5 \text{ A} + 1 \text{ A} + 2 \text{ A}$
Answer: $I_T = 3.5$ A
$\qquad I_T = I_1 = I_2 = 3.5$ A

Formula: $I_4 = I_6 + I_7$
Substitution: $I_4 = 1 \text{ A} + 2 \text{ A}$
Answer: $I_4 = 3$ A
$\qquad I_4 = I_5 = 3$ A

$I_1 = 3.5$ A	$I_5 = 3$ A
$I_2 = 3.5$ A	$I_6 = 1$ A
$I_3 = 0.5$ A	$I_7 = 2$ A
$I_4 = 3$ A	$I_8 = 2$ A

Chapter Test

7

Name: _____

Date: _____

Class: _____

Parallel Circuits

1. Describe these characteristics of a parallel circuit.

   a. Branch current: _____

   b. Mainline current: _____

   c. Voltage across parallel branches: _____

2. Write three formulas used to calculate total resistance in parallel, not including Ohm's law or power formulas.

   _____

   _____

   _____

_____ 3. Total resistance in a parallel circuit is:
   a. the sum of the branch resistances.
   b. smaller than the smallest branch resistance.
   c. larger than the largest branch resistance.
   d. the average of the branch resistances.

_____ 4. Total current in a parallel circuit is:
   a. smaller than the smallest branch current.
   b. is the same at all points in the circuit.
   c. the sum of the branch currents.
   d. twice the mainline current.

_____ 5. The shortcut formula can be used to solve for total resistance in a parallel circuit when:
   a. several resistors have different values.
   b. the sum of the branch resistances does not equal the total.
   c. the reciprocals have too many decimal places.
   d. there are only two resistors in parallel.

_____ 6. The equal-value resistors formula can be used to solve for total resistance in a parallel circuit when:
   a. several resistors have the same value.
   b. the sum of the branch resistances does not equal the total.
   c. the reciprocals have too many decimal places.
   d. there are only two resistors in parallel.

_____   7. In a parallel circuit, if one branch is shorted:
        a. the fuse will blow.
        b. there will be no effect on the other branches.
        c. total resistance will be increased.
        d. total current will be reduced by the missing branch.

_____   8. In a parallel circuit, if one branch has an open:
        a. the fuse will blow.
        b. all circuit current will be stopped.
        c. total resistance will equal zero.
        d. there will be no effect on the other branches.

With each problem, write the formula, substitution, and answer.

9. Using the information given in the schematic diagram and the reciprocal (conductance) formula, find the value of total resistance.

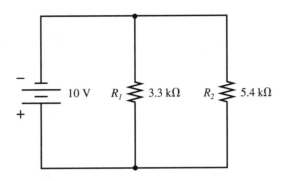

Formula: _____

Substitution: _____

Answer: _____

10. Find the total resistance of the circuit in question 9 using the shortcut formula.

Formula: _____

Substitution: _____

Answer: _____

11. Find the total resistance of the circuit shown.

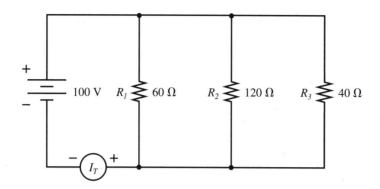

Formula: _____

Substitution: _____

Answer: _____

12. Find the total current of the circuit in question 11.

Formula: _____

Substitution: _____

Answer: _____

13. Find the total resistance of the circuit shown, using the reciprocal (conductance) formula.

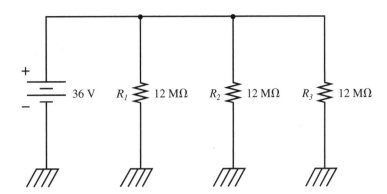

Formula: _____

Substitution: _____

Answer: _____

14. Find the total resistance of the circuit in question 13 using the equal-value resistor formula.

Formula: _____

Substitution: _____

Answer: _____

15. Find the applied voltage of the circuit shown, using your choice of formula.

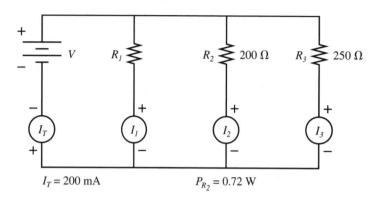

$I_T = 200$ mA $P_{R_2} = 0.72$ W

Formula: _____

Substitution: _____

Answer: _____

16. Find the total resistance of the circuit shown, using your choice of formula.

Formula: _____

Substitution: _____

Answer: _____

17. Find the total resistance of the circuit shown.

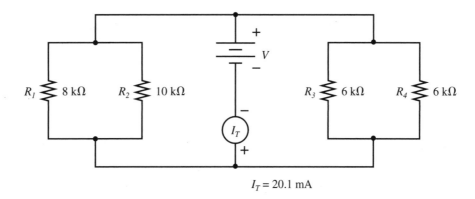

$I_T = 20.1$ mA

Formula: _____

Substitution: _____

Answer: _____

18. Find the applied voltage for the circuit for question 17.

Formula: _____

Substitution: _____

Answer: _____

19. Find the total power for the circuit for question 17.

Formula: _____

Substitution: _____

Answer: _____

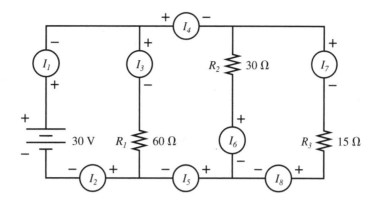

Name: _____

20. Calculate the current measured in each of the ammeters shown in the circuit. Place the answers on the lines.

$I_1 =$ _____

$I_2 =$ _____

$I_3 =$ _____

$I_4 =$ _____

$I_5 =$ _____

$I_6 =$ _____

$I_7 =$ _____

$I_8 =$ _____

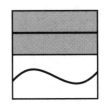

Series-Parallel Circuits

OBJECTIVES

After studying this chapter, students should be able to:
- Differentiate between series connected resistors and parallel connected resistors.
- Apply the rules and circuit formulas to combination circuits.
- Calculate the circuit parameters of increasingly difficult combinations.
- Examine the range resistors connected internally in a voltmeter, ammeter, and ohmmeter.

INSTRUCTIONAL MATERIALS

Text: Pages 211–240
Test Your Knowledge Questions, Pages 236–239
Study Guide: Pages 65–78
Laboratory Manual: Pages 71–75

ANSWERS TO TEXTBOOK

Test Your Knowledge, Pages 236–239
1. Find the equivalent series resistance for each parallel branch. Then, solve the parallel circuit.
2. A circuit is easier to visualize when drawn as an equivalent circuit. It is also easier to find the circuit parameters.
3. It drops the voltage.
4. When the voltmeter is connected across a very high resistance.
5. If the resistance is too high for an analog meter, using digital meter can solve the problem.
6. The ohms per volt rating is multiplied by the maximum voltage of the scale in question.
7. Shunt resistors are connected in parallel with the meter movement of an ammeter.
8. It bypasses some of the current around the meter movement to change the meter's range.
9. A multiplier resistor is connected in series with the meter movement of a voltmeter.
10. It drops some of the voltage to change the meter's range.

11. Branch resistance $R_A = 10\ \Omega$
Branch resistance $R_B = 2\ \text{k}\Omega$
Branch resistance $R_C = 1\ \text{k}\Omega$
Branch current $I_A = 1\ \text{A}$
Branch current $I_B = 5\ \text{mA}$
Branch current $I_C = 10\ \text{mA}$
Total current $I_T = 1.015\ \text{A}$
Total resistance $R_T = 9.85\ \Omega$
Voltage drop across R_1, $V_{R_1} = 5\ \text{V}$
Voltage drop across R_2, $V_{R_2} = 5\ \text{V}$
Voltage drop across R_3, $V_{R_3} = 10\ \text{V}$
Voltage drop across R_4, $V_{R_4} = 1\ \text{V}$
Voltage drop across R_5, $V_{R_5} = 5\ \text{V}$
Voltage drop across R_6, $V_{R_6} = 4\ \text{V}$

12. Although not asked for in the problem, it is necessary to first find the equivalent resistance of each parallel combination.

$R_A = 15\ \Omega$

$R_B = 10\ \Omega$

$R_D = 6\ \Omega$

$R_E = 10\ \Omega$

Total resistance, $R_T = 131\ \Omega$

Total current, $I_T = 1\ \text{A}$

Voltage drop, $V_{R_1} = 30\ \text{V}$

Voltage drop, $V_{R_A} = 15\ \text{V}$

Voltage drop across R_2 to $R_9 = 7.5\ \text{V}$ (each)

Voltage drop $V_{R_B} = 10\ \text{V}$

Voltage drop across R_{10} to $R_{12} = 10\ \text{V}$ (each)

Voltage drop, $V_{R_{13}}$ = 30 V

Voltage drop, $V_{R_{14}}$ = 30 V

Voltage drop, V_{R_D} = 6 V

    Voltage drop across R_{15} to R_{19} = 6 V (each)

Voltage drop, V_{R_E} = 10 V

    Voltage drop across R_{20} and R_{23} = 10 V

    Voltage drop across R_{21}, R_{22}, R_{24}, and R_{25} = 5 V

13. R_T = 20 Ω

    I_T = 2.5 A

    I_{R_1} = 2.5 A

    I_{R_2} = 1.25 A

    I_{R_3} = 2.5 A

    I_{R_4} = 1.25 A

    I_{R_5} = 1.25 A

    I_{R_6} = 1.25 A

    V_{R_1} = 12.5 V

    V_{R_2} = 25 V

    V_{R_3} = 12.5 V

    V_{R_4} = 6.25 V

    V_{R_5} = 12.5 V

    V_{R_6} = 6.25 V

14. R_T = 100 Ω

    I_T = 1 A

    I_{R_1} = 1 A

    I_{R_2} = 0.5 A

    I_{R_3} = 1 A

    I_{R_4} = 0.5 A

    I_{R_5} = 0.25 A

    I_{R_6} = 0.5 A

    I_{R_7} = 0.25 A

    I_{R_8} = 0.25 A

    I_{R_9} = 0.25 A

    V_{R_1} = 20 V

    V_{R_2} = 50 V

    V_{R_3} = 30 V

    V_{R_4} = 25 V

    V_{R_5} = 12.5 V

    V_{R_6} = 12.5 V

    V_{R_7} = 3.75 V

    V_{R_8} = 2.5 V

    V_{R_9} = 6.25 V

15. Circuit resistance without meter = 1 MΩ

    Current without meter = 50 μA

    Voltage across R_2 without meter = 25 V

    Meter input resistance = 250 kΩ

    Combination of R_2 and meter = 167 kΩ

    Circuit resistance with meter = 667 kΩ

    Current with meter = 75 μA

    Voltage across R_2 with meter = 12.5 V

16. Load current = 8 A

    Load resistance = 0.75 Ω

        R_2 = 0.75 Ω and I_{R_2} = 8 A

    Combination of R_2 and load = 0.375 Ω and 16 A

        V_{R_1} = 6 V and I_{R_1} = 16 A

        R_1 = 0.375 Ω

ANSWERS TO STUDY GUIDE

Pages 65–78

  1. a.

  2. b.

  3. c.

  4. b.

  5. a.

  6. d.

  7. a.

  8. d.

  9. a.

10. b.

11. a. Branch resistance $R_A = R_1$ = 50 Ω

      b. Branch resistance R_B:
          Formula: $R_B = R_2 + R_3$
          Substitution: R_B = 175 Ω + 75 Ω
          Answer: R_B = 250 Ω

      c. Branch resistance R_C:
          Formula: $R_C = R_4 + R_5 + R_6$
          Substitution: R_C = 300 Ω + 500 Ω + 200 Ω
          Answer: R_C = 1000 Ω

d. Branch current I_A:

Formula: $I_A = \dfrac{V}{R_A}$

Substitution: $I_A = \dfrac{25 \text{ V}}{50 \text{ }\Omega}$

Answer: $I_A = 0.5$ A

e. Branch current I_B:

Formula: $I_B = \dfrac{V}{R_B}$

Substitution: $I_B = \dfrac{25 \text{ V}}{250 \text{ }\Omega}$

Answer: $I_B = 0.1$ A

f. Branch current I_C:

Formula: $I_C = \dfrac{V}{R_C}$

Substitution: $I_C = \dfrac{25 \text{ V}}{1000 \text{ }\Omega}$

Answer: $I_C = 0.025$ A

g. Total resistance:

Formula: $\dfrac{1}{R_T} = \dfrac{1}{R_1} + \dfrac{1}{R_2} + \dfrac{1}{R_3}$

Substitution:

$$\dfrac{1}{R_T} = \dfrac{1}{50 \text{ }\Omega} + \dfrac{1}{250 \text{ }\Omega} + \dfrac{1}{1000 \text{ }\Omega}$$

Answer: $R_T = 40 \text{ }\Omega$

h. Voltage drop, V_1:
Formula: $V_1 = I_A \times R_1$
Substitution: $V_1 = 0.5$ A $\times 50 \text{ }\Omega$
Answer: $V_1 = 25$ V

i. Voltage drop, $V_2 = 17.5$ V

j. Voltage drop, $V_3 = 7.5$ V

k. Voltage drop, $V_4 = 7.5$ V

l. Voltage drop, $V_5 = 12.5$ V

m. Voltage drop, $V_6 = 5$ V

12. a. Resistance of group A:

Formula: $\dfrac{1}{R_A} = \dfrac{1}{R_1} + \dfrac{1}{R_2 + R_3} + \dfrac{1}{R_4 + R_5} + \dfrac{1}{R_6}$

Substitution: $\dfrac{1}{R_A} = \dfrac{1}{50 \text{ }\Omega} + \dfrac{1}{50 \text{ }\Omega + 50 \text{ }\Omega} +$

$$\dfrac{1}{50 \text{ }\Omega + 50 \text{ }\Omega} + \dfrac{1}{50 \text{ }\Omega}$$

Answer: $R_A = 16.7 \text{ }\Omega$

b. Answer: $R_B = 12.5 \text{ }\Omega$

c. Answer: $R_C = 50 \text{ }\Omega$

d. Answer: $R_D = 10 \text{ }\Omega$

e. Answer: $R_E = 50 \text{ }\Omega$

f. Answer: $R_F = 16.7 \text{ }\Omega$

g. Answer: $R_G = 50 \text{ }\Omega$

h. Student sketch of equivalent circuit.

i. Total resistance R_T:
Formula:
$$R_T = R_A + R_B + R_C + R_D + R_E + R_F + R_G$$
Substitution: $R_T = 16.7 \text{ }\Omega + 12.5 \text{ }\Omega + 50 \text{ }\Omega$
$+ 10 \text{ }\Omega + 50 \text{ }\Omega + 16.7 \text{ }\Omega + 50 \text{ }\Omega$
Answer: $R_T = 206 \text{ }\Omega$

j. Total current I_T:

Formula: $I_T = \dfrac{V}{R_T}$

Substitution: $I_T = \dfrac{100 \text{ V}}{206 \text{ }\Omega}$

Answer: $I_T = 0.485$ A

k. Voltage drop V_{R_A}:
    Formula: $V_{R_A} = I_T \times R_A$
    Substitution: $V_{R_A} = 0.485 \text{ A} \times 16.7 \text{ } \Omega$
    Answer: $V_{R_A} = 8.1 \text{ V}$

l. Voltage drop $V_{R_B} = 6.06 \text{ V}$

m. Voltage drop $V_{R_C} = 24.25 \text{ V}$

n. Voltage drop $V_{R_D} = 4.85 \text{ V}$

o. Voltage drop $V_{R_E} = 24.25 \text{ V}$

p. Voltage drop $V_{R_F} = 8.1 \text{ V}$

q. Voltage drop $V_{R_G} = 24.25 \text{ V}$

13. a. Series combination of R_4, R_5, R_6:
     Formula: $R_{4\text{-}5\text{-}6} = R_4 + R_5 + R_6$
     Substitution: $R_{4\text{-}5\text{-}6} = 30 \text{ } \Omega + 40 \text{ } \Omega + 20 \text{ } \Omega$
     Answer: $R_{4\text{-}5\text{-}6} = 90 \text{ } \Omega$

     b. Student sketch of equivalent circuit.

     c. Parallel combination of R_3 with $R_{4\text{-}5\text{-}6}$:

     Formula: $\dfrac{1}{R_{3\text{-}4\text{-}5\text{-}6}} = \dfrac{1}{R_3} + \dfrac{1}{R_{4\text{-}5\text{-}6}}$

     Answer: $R_{3\text{-}4\text{-}5\text{-}6} = 45 \text{ } \Omega$

     d. Student sketch of equivalent circuit.

     e. Series combination of R_1, $R_{3\text{-}4\text{-}5\text{-}6}$, R_2:
     Formula $R_T = R_1 + R_{3\text{-}4\text{-}5\text{-}6}, + R_3$
     Substitution: $R_T = 20 \text{ } \Omega + 45 \text{ } \Omega + 35 \text{ } \Omega$
     Answer: $R_T = 100 \text{ } \Omega$

     f. Student sketch of equivalent circuit.

g. Total current:

    Formula: $I_T = \dfrac{V}{R_T}$

    Substitution: $I_T = \dfrac{100 \text{ V}}{100 \text{ } \Omega}$

    Answer: $I_T = 1 \text{ A}$

h. $I_{R_1} = 1 \text{ A}$
    $I_{R_2} = 1 \text{ A}$
    $I_{R_{3\text{-}4\text{-}5\text{-}6}} = 0.5 \text{ A}$

i. Voltage across R_1:
    Formula: $V_{R_1} = I_{R_1} \times R_1$
    Substitution: $V_{R_1} = 1 \text{ A} \times 20 \text{ } \Omega$
    Answer: $V_{R_1} = 20 \text{ V}$

j. Answer: $V_{R_2} = 35 \text{ V}$

k. Answer: $V_{R_{3\text{-}4\text{-}5\text{-}6}} = 45 \text{ V}$

l. Answer: $V_{R_3} = 45 \text{ V}$

m. Current through resistors 4, 5, 6:

    Formula: $I_{R_{4\text{-}5\text{-}6}} = \dfrac{V_{R_{4\text{-}5\text{-}6}}}{R_{4\text{-}5\text{-}6}}$

    Substitution: $I_{R_{4\text{-}5\text{-}6}} = \dfrac{45 \text{ V}}{90 \text{ } \Omega}$

    Answer: $I_{R_{4\text{-}5\text{-}6}} = 0.5 \text{ A}$

n. Answer: $V_{R_4} = 15 \text{ V}$

o. Answer: $V_{R_5} = 20 \text{ V}$

p. Answer: $V_{R_6} = 10 \text{ V}$

14. a. Series combination of R_7, R_8, R_9:
     Formula: $R_{7\text{-}8\text{-}9} = R_7 + R_8 + R_9$
     Substitution: $R_{7\text{-}8\text{-}9} = 100 \text{ } \Omega + 200 \text{ } \Omega + 300 \text{ } \Omega$
     Answer: $R_{7\text{-}8\text{-}9} = 600 \text{ } \Omega$

b.   Student sketch of equivalent circuit.

c.   Parallel combination of R_5 with $R_{7\text{-}8\text{-}9}$:

Formula: $\dfrac{1}{R_{5\text{-}7\text{-}8\text{-}9}} = \dfrac{1}{R_5} + \dfrac{1}{R_{7\text{-}8\text{-}9}}$

Substitution: $\dfrac{1}{R_{5\text{-}7\text{-}8\text{-}9}} = \dfrac{1}{600 \ \Omega} + \dfrac{1}{600 \ \Omega}$

Answer: $R_{5\text{-}7\text{-}8\text{-}9} = 300 \ \Omega$

d.   Student sketch of equivalent circuit.

e.   Series combination of R_4, R_6 and $R_{5\text{-}7\text{-}8\text{-}9}$:
Formula: $R_{4\text{-}5\text{-}6\text{-}7\text{-}8\text{-}9} = R_4 + R_6 + R_{5\text{-}7\text{-}8\text{-}9}$
Substitution:
$\quad R_{4\text{-}5\text{-}6\text{-}7\text{-}8\text{-}9} = 150 \ \Omega + 150 \ \Omega + 300 \ \Omega$
Answer: $R_{4\text{-}5\text{-}6\text{-}7\text{-}8\text{-}9} = 600 \ \Omega$

f.   Student sketch of equivalent circuit.

g.   Parallel combination of R_2 with $R_{4\text{-}5\text{-}6\text{-}7\text{-}8\text{-}9}$:

Formula: $\dfrac{1}{R_{2\text{-}4\text{-}5\text{-}6\text{-}7\text{-}8\text{-}9}} = \dfrac{1}{R_2} + \dfrac{1}{R_{4\text{-}5\text{-}6\text{-}7\text{-}8\text{-}9}}$

Substitution: $\dfrac{1}{R_{2\text{-}4\text{-}5\text{-}6\text{-}7\text{-}8\text{-}9}} = \dfrac{1}{600 \ \Omega} + \dfrac{1}{600 \ \Omega}$

Answer: $R_{2\text{-}4\text{-}5\text{-}6\text{-}7\text{-}8\text{-}9} = 300 \ \Omega$

h.   Student sketch of equivalent circuit.

i.   Series combination of R_1, R_3 and $R_{2\text{-}4\text{-}5\text{-}6\text{-}7\text{-}8\text{-}9}$:
Formula: $R_T = R_1 + R_3 + R_{2\text{-}4\text{-}5\text{-}7\text{-}8\text{-}9}$
Substitution: $R_T = 100 \ \Omega + 200 \ \Omega + 300 \ \Omega$
Answer: $R_T = 600 \ \Omega$

j.   Student sketch of equivalent circuit.

15. a.   Total resistance without meter:
Formula: $R_T = R_1 + R_2 + R_3$
Substitution:
$\quad R_T = 250 \ \text{k}\Omega + 300 \ \text{k}\Omega + 450 \ \text{k}\Omega$
Answer: $R_T = 1 \ \text{M}\Omega$

b.   Current without meter:

Formula: $I = \dfrac{V}{R_T}$

Substitution: $I = \dfrac{20 \ \text{V}}{1 \ \text{M}\Omega}$

Answer: $I = 20 \ \mu\text{A}$

c.   Voltage across R_1 without meter:
Formula: $V_{R_1} = I \times R_1$
Substitution: $V_{R_1} = 20 \ \mu\text{A} \times 250 \ \text{k}\Omega$
Answer: $V_{R_1} = 5 \ \text{V}$

d.   Voltage across R_2 without meter:
Answer: $R_2 = 6 \ \text{V}$

e.   Voltage across R_3 without meter = 9 V
Answer: $R_3 = 6 \ \text{V}$

f.   Meter input resistance: 150 kΩ

g.   Total resistance with meter:
Formula: $R_T = R_1 + R_2 \parallel R_M + R_3$
Substitution:
$\quad R_T = 250 \ \text{k}\Omega + 100 \ \text{k}\Omega + 450 \ \text{k}\Omega$
Answer: $R_T = 800 \ \text{k}\Omega$

h.   Current with meter:

Formula: $I = \dfrac{V}{R_T}$

Substitution: $I = \dfrac{20 \ \text{V}}{800 \ \text{k}\Omega}$

Answer: $I = 25 \ \mu\text{A}$

i.   Voltage across R_1 with meter:
Formula: $V_{R_1} = I \times R_1$
Substitution: $V_{R_1} = 25 \ \mu\text{A} \times 250 \ \text{k}\Omega$
Answer: $V_{R_1} = 6.25 \ \text{V}$

j.   Voltage across R_2 with meter = 2.5 V

k.   Voltage across R_3 with meter = 11.25 V

16. a.   Resistance of the load:

Formula: $R_L = \dfrac{V^2}{P}$

Substitution: $R_L = \dfrac{6^2 \text{ V}}{36 \text{ W}}$

Answer: $R_L = 1 \ \Omega$

  b: Load current:

Formula: $I_L = \dfrac{V_L}{R_L}$

Substitution: $I_L = \dfrac{6 \text{ V}}{1 \ \Omega}$

Answer: $I_L = 6$ A

  c. $R_2 = R_L = 1 \ \Omega$

$I_{R_2} = I_L = 6$ A

Formula: $I_T = I_L + I_{R_2}$

Substitution: $I_T = 6$ A $+ 6$ A

Answer: $I_T = 12$ A

  d.   Voltage across R_1:

Formula: $V_{R_1} =$ combination of R_2 and
load $\times I_T$

Substitution: $V_{R_1} = 0.5 \ \Omega \times 12$ A

Answer: $V_{R_1} = 6$ V

  e.   Value of R_1:

Formula: $R_1 = \dfrac{V_{R_1}}{I_{R_1}}$

Substitution: $R_1 = \dfrac{6 \text{ V}}{12 \text{ A}}$

Answer: $R_1 = 0.5 \ \Omega$

ANSWERS TO CHAPTER TEST IN THE INSTRUCTOR'S MANUAL

Pages 105–118

1. b.
2. a.
3. d.
4. a.

5. d.

6. a.   Resistance of group A:

Formula:

$$\frac{1}{R_A} = \frac{1}{R_1} + \frac{1}{R_2 + R_3} + \frac{1}{R_4 + R_5} + \frac{1}{R_6}$$

Substitution:

$$\frac{1}{R_A} = \frac{1}{100 \ \Omega} + \frac{1}{200 \ \Omega} + \frac{1}{200 \ \Omega} + \frac{1}{100 \ \Omega}$$

Answer: $R_A = 33.3 \ \Omega$

  b.   Resistance of group B:

Formula: $R_B = \dfrac{R}{N}$

Substitution: $R_B = \dfrac{100 \ \Omega}{4}$

Answer: $R_B = 25 \ \Omega$

  c.   Resistance of group C:

$R_C = R_{11} = 100 \ \Omega$ (series resistor with no
parallel branches)

  d.   Resistance of group D:

Formula: $R_D = \dfrac{R}{N}$

Substitution: $R_D = \dfrac{100 \ \Omega}{5}$

Answer: $R_D = 20 \ \Omega$

  e.   Resistance of group E:

Formula:

$$\frac{1}{R_E} = \frac{1}{R_{17} + R_{18}} + \frac{1}{R_{19}} + \frac{1}{R_{20} + R_{21}} + \frac{1}{R_{22}}$$

Substitution:

$$\frac{1}{R_E} = \frac{1}{200 \ \Omega} + \frac{1}{100 \ \Omega} + \frac{1}{200 \ \Omega} + \frac{1}{100 \ \Omega}$$

Answer: $R_E = 33.3 \ \Omega$

f.   Draw the equivalent circuit.

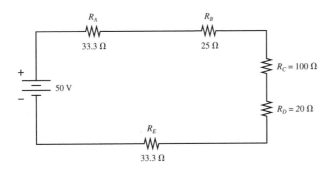

7.   a.   Total resistance:
Formula: $R_T = R_A + R_B + R_C + R_D + R_E$
Substitution: $R_T = 33.3 \ \Omega + 25 \ \Omega + 100 \ \Omega$
$+ \ 20 \ \Omega + 33.3 \ \Omega$
Answer: $R_T = 212 \ \Omega$

b.   Total current:

Formula: $I_T = \dfrac{V}{R_T}$

Substitution: $I_T = \dfrac{50 \ \text{V}}{212 \ \Omega}$

Answer: $I_T = 0.236 \ \text{A}$

c.   Voltage drop across group B:
Formula: $V_{R_B} = I_T \times R_B$
Substitution: $V_{R_B} = 0.236 \ \text{A} \times 25 \ \Omega$
Answer: $V_{R_B} = 5.9 \ \text{V}$

8.   a.   Combination of resistors 4, 5, and 6:
Formula: $R_{4\text{-}5\text{-}6} = R_4 + R_5 + R_6$
Substitution: $R_{4\text{-}5\text{-}6} = 30 \ \Omega + 20 \ \Omega + 10 \ \Omega$
Answer: $R_{4\text{-}5\text{-}6} = 60 \ \Omega$

b.   Draw an equivalent circuit:

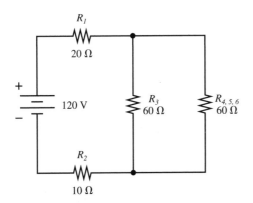

c.   Combination of resistor 3 with equivalent resistor 4-5-6:

Formula: $R_{3\text{-}4\text{-}5\text{-}6} = \dfrac{R}{N}$

Substitution: $R_{3\text{-}4\text{-}5\text{-}6} = \dfrac{60 \ \Omega}{2}$

Answer: $R_{3\text{-}4\text{-}5\text{-}6} = 30 \ \Omega$

d.   Draw an equivalent circuit:

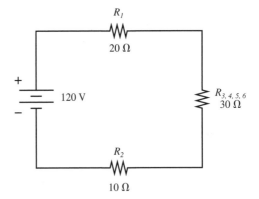

e.   Combination of resistors 1 and 2 with 3-4-5-6:
Formula: $R_{1\text{-}2\text{-}3\text{-}4\text{-}5\text{-}6} = R_1 + R_2 + R_{3\text{-}4\text{-}5\text{-}6}$
Substitution: $R_{1\text{-}2\text{-}3\text{-}4\text{-}5\text{-}6} = 20 \ \Omega + 10 \ \Omega + 30 \ \Omega$
Answer: $R_{1\text{-}2\text{-}3\text{-}4\text{-}5\text{-}6} = 60 \ \Omega$

f.   Draw an equivalent circuit:

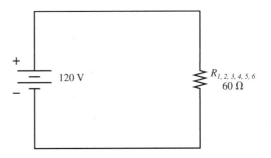

9.   a.   Using the circuit in step f, find the total current.

Formula: $I_T = \dfrac{V}{R_{1\text{-}2\text{-}3\text{-}4\text{-}5\text{-}6}}$

Substitution: $I_T = \dfrac{120 \ \text{V}}{60 \ \Omega}$

Answer: $I_T = 2 \ \text{A}$

b. Use the circuit in step d to find:
Current through R_1: $I_{R_1} = I_T = 2$ A
Current through R_2: $I_{R_2} = I_T = 2$ A
Current through $R_{3\text{-}4\text{-}5\text{-}6}$: $I_{R_{3\text{-}4\text{-}5\text{-}6}} = I_T = 2$ A

c. Voltage across $R_{3\text{-}4\text{-}5\text{-}6}$, using the circuit in step d:
Formula: $V_{R_{3\text{-}4\text{-}5\text{-}6}} = I_T \times R_{3\text{-}4\text{-}5\text{-}6}$
Substitution: $V_{R_{3\text{-}4\text{-}5\text{-}6}} = 2$ A $\times$ 30 Ω
Answer: $V_{R_{3\text{-}4\text{-}5\text{-}6}} = 60$ V

d. Current through resistors 4, 5, and 6, using the circuit in step b:

Formula: $I_{R_{4\text{-}5\text{-}6}} = \dfrac{V_{R_{4\text{-}5\text{-}6}}}{R_{4\text{-}5\text{-}6}}$

Substitution: $I_{R_{4\text{-}5\text{-}6}} = \dfrac{60 \text{ V}}{60 \text{ }\Omega}$

Answer: $I_{R_{4\text{-}5\text{-}6}} = 1$ A

e. Voltage across R_6, using the original:
Formula: $V_{R_6} = I_{R_{4\text{-}5\text{-}6}} \times R_6$
Substitution: $V_{R_6} = 1$ A $\times$ 10 Ω
Answer: $V_{R_6} = 10$ V

d. Voltage across R_1:
Formula: $V_{R_1} = V - V_{R_2}$
Substitution: $V_{R_1} = 36$ V $-$ 12 V
Answer: $V_{R_1} = 24$ V

e. Value of R_1:
Formula: $R_1 = \dfrac{V_{R_1}}{I_{R_1}}$

Substitution: $R_1 = \dfrac{24 \text{ V}}{0.8 \text{ A}}$

Answer: $R_1 = 30$ Ω

10. a. Resistance of the load and R_2:

Formula: $R_L = \dfrac{V^2}{P_L}$

Substitution: $R_L = \dfrac{(12 \text{ V})^2}{4.8 \text{ W}}$

Answer: $R_L = 30$ Ω

b. Load current and current in R_2:

Formula: $I_L = \dfrac{P_L}{V_L}$

Substitution: $I_L = \dfrac{4.8 \text{ W}}{12 \text{ V}}$

Answer: $I_L = 0.4$ A

c. Total current:
Formula: $I_T = I_L + I_2$
Substitution: $I_T = 0.4$ A $+$ 0.4 A
Answer: $I_T = 0.8$ A

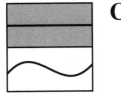

Chapter Test

8

Series-Parallel Circuits

Select the best answer.

_____ 1. Voltmeter loading has the greatest effect when connected to a:
   a. parallel circuit with three equal-value resistors.
   b. series circuit with large resistance values.
   c. series circuit with small resistance values.
   d. circuit with a blown fuse.

_____ 2. To avoid voltmeter loading, the internal resistance should be:
   a. at least 10 times larger than the circuit resistance.
   b. temporarily disconnected.
   c. equal to the circuit resistance.
   d. considered zero ohms.

_____ 3. When calculating the input resistance of a voltmeter, the ohms per volt rating is:
   a. the resistance for each volt in the circuit.
   b. a measure of the voltage drop of the meter.
   c. used to calculate maximum meter current.
   d. multiplied times the full scale voltage.

_____ 4. A shunt resistor is connected in _____ with the meter movement of _____.
   a. parallel, an ammeter
   b. series, an ammeter
   c. parallel, a voltmeter
   d. series, a voltmeter

_____ 5. A multiplier resistor is connected in _____ with the meter movement of _____.
   a. parallel, an ammeter
   b. series, an ammeter
   c. parallel, a voltmeter
   d. series, a voltmeter

With each problem, write the formula, substitution, and answer.

6. For the circuit shown in the following figure, find the resistance of each group. Also, draw an equivalent circuit.

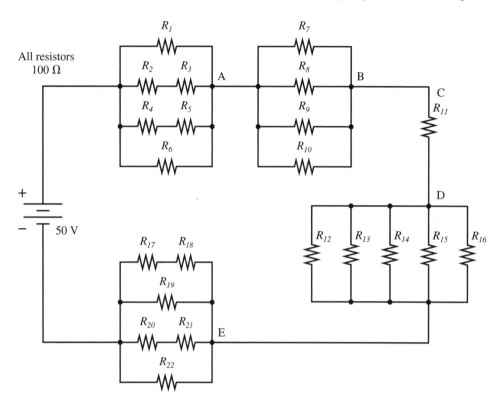

a. Resistance of group A:

Formula: _____

Substitution: _____

Answer: _____

b. Resistance of group B:

Formula: _____

Substitution: _____

Answer: _____

c. Resistance of group C:

Formula: _____

Substitution: _____

Answer: _____

d. Resistance of group D:

Formula: _____

Substitution: _____

Answer: _____

e. Resistance of group E:

Formula: _____

Substitution: _____

Answer: _____

f. Draw the equivalent circuit in the space that follows.

7. Use the equivalent circuit in question 6 to find these values.

a. Total resistance:

Formula: _____

Substitution: _____

Answer: _____

b. Total current:

Formula: _____

Substitution: _____

Answer: _____

c. Voltage drop across group B:

Formula: _____

Substitution: _____

Answer: _____

8. For the combination circuit shown in the figure below, calculate the total resistance. Draw an equivalent circuit for each step.

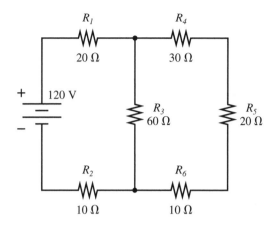

a. Combination of resistors 4, 5, and 6:

Formula: _____

Substitution: _____

Answer: _____

b. Draw an equivalent circuit:

c. Combination of resistor 3 with equivalent resistor 4-5-6:

Formula: _____

Substitution: _____

Answer: _____

d. Draw an equivalent circuit:

e. Combination of resistors 1 and 2 with 3-4-5-6:

Formula: _____

Substitution: _____

Answer: _____ (Total resistance)

f. Draw an equivalent circuit:

9. Use the equivalent circuits found in question 8.

    a. Using the circuit in step f, find the total current.

        Formula: _____

        Substitution: _____

        Answer: _____

    b. Use the circuit in step d to find:

        Current through R_1: _____

        Current through R_2: _____

        Current through $R_{3\text{-}4\text{-}5\text{-}6}$: _____

    c. Find the voltage across $R_{3\text{-}4\text{-}5\text{-}6}$, using the circuit in step d.

        Formula: _____

        Substitution: _____

        Answer: _____

Name: ____________________________

d. Find the current through resistors 4, 5, and 6, using the circuit in step b.

Formula: ____

Substitution: ____

Answer: ____

e. Find the voltage across R_6, using the original circuit.

Formula: ____

Substitution: ____

Answer: ____

10. Calculate the values of R_1 and R_2 in the voltage divider circuit. The value of R_2 should equal the load.

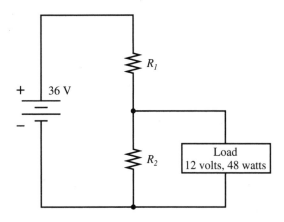

a. Resistance of the load and R_2:

Formula: _____

Substitution: _____

Answer: _____

b. Load current and current in R_2:

Formula: _____

Substitution: _____

Answer: _____

c. Total current:

Formula: _____

Substitution: _____

Answer: _____

d. Voltage across R_1:

Formula: _____

Substitution: _____

Answer: _____

e. Value of R_1:

Formula: _____

Substitution: _____

Answer: _____

Chapter 9

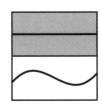
DC Circuit Theorems

OBJECTIVES

After studying this chapter, students should be able to:
- Solve dc circuits using Kirchhoff's voltage and current laws, the superposition theorem, Thevenin's theorem, and Norton's theorem.
- Solve dc circuits with more than one voltage source using Kirchhoff's laws and the superposition theorem.
- Calculate the values of a Wheatstone bridge.

INSTRUCTIONAL MATERIALS

Text: Pages 241–281
Test Your Knowledge Questions, Pages 277–281
Study Guide: Pages 79–92
Laboratory Manual: Pages 77–93

ANSWERS TO TEXTBOOK

Test Your Knowledge, Pages 277–281

1. Current law: The algebraic sum of the currents entering and leaving a point will equal zero.
2. Voltage law: The algebraic sum of the voltages around a loop will equal zero.
3. Write the loop equation by starting at one point and continuing around the loop in the assumed direction of current flow. The positive and negative sign of the voltage is taken from the sign of the polarity as the current first enters the component.
4. In a circuit containing more than one voltage source, the current or voltage of the individual components is the algebraic sum of the sources acting separately.
5. The Thevenin voltage is found by removing the load resistor and finding what voltage would appear at the load terminals.

6. The Thevenin resistance is measured at the load terminals with the voltage removed and replaced by a short.
7. Student drawing of Thevenin circuits will vary.
8. Find load voltage and current by treating the Thevenin equivalent circuit as a series circuit.
9. R_{TH} and V_{TH} remain the same. A new load will change the current and load voltage.
10. Norton current is found by replacing the load with a short.
11. Norton resistance is measured at the load terminals with the voltage replaced by a short.
12. Student drawing of Norton circuits will vary.
13. Norton and Thevenin resistances are the same.
14. $I_2 = 60$ mA

15. Junction D: $I_7 = I_5 + I_6 = 15$ A
Mainline current: $I_1 = I_7 = 15$ A
Junction A: $I_2 = I_1 - I_3 = 5$ A
Junction B: $I_4 = I_2 - I_5 = -3$ A (I_4 flows into junction B)
Junction C: $I_4 = I_3 - I_6 = 3$ A (I_4 flows away from junction C)

16. $V - V_{R_3} - V_{R_2} - V_{R_1} = 0$
$V_{R_2} = 25$ V

17. a. $V - V_{R_5} - V_{R_2} - V_{R_1} = 0$
$V_{R_2} = 40$ V

b. $V_{R_2} - V_{R_3} - V_{R_4} = 0$
$V_{R_3} = 25$ V

18. Loop A:
Equation: $I_A R_1 + (I_A + I_B)R_3 - V_A = 0$
Substitution: $35I_A + 15(I_A + I_B) = 25$
Simplification: $35I_A + 15I_A + 15I_B = 25$
$50I_A + 15I_B = 25$

Loop B:

Equation: $I_B R_2 + (I_A + I_B)R_3 - V_B = 0$

Substitution: $25I_B + 15(I_A + I_B) = 40$

Simplification: $25I_B + 15I_A + 15I_B = 40$

$$15I_A + 40I_B = 40$$

Combining:

Multiply equation A by 8:

$$400I_A + 120I_B = 200$$

Multiply equation B by 3:

$$45I_A + 120I_B = 120$$

Subtracting the equations: $355I_A = 80$

$$I_A = 0.225 \text{ A}$$

Substituting: $15I_A + 40I_B = 40$

$$15(0.225 \text{ A}) + 40I_B = 40$$

$$I_B = 0.915 \text{ A}$$

Answers:

$I_{R_1} = 0.225$ A

$V_{R_1} = 7.88$ V

$I_{R_2} = 0.915$ A

$V_{R_2} = 22.9$ V

$I_{R_3} = 1.14$ A

$V_{R_3} = 17.1$ V

19. V_B replaced with a short:

$R_{T_A} = 25.5\ \Omega$

$I_{T_A} = 5.88$ A

$V_{R_{1_A}} = 88.2$ V

$I_{R_{1_A}} = 5.88$ A

$V_{R_{2_A}} = 61.8$ V

$I_{R_{2_A}} = 1.77$ A

$V_{R_{3_A}} = 61.8$ V

$I_{R_{3_A}} = 4.12$ A

V_A replaced with a short:

$R_{T_B} = 42.5\ \Omega$

$I_{T_B} = 2.35$ A

$V_{R_{1_B}} = 17.7$ V

$I_{R_{1_B}} = 1.18$ A

$V_{R_{2_B}} = 82.3$ V

$I_{R_{2_B}} = 2.35$ A

$V_{R_{3_B}} = 17.7$ V

$I_{R_{3_B}} = 1.18$ A

Combining:

$V_{R_1} = -88.2 + 17.7 = 70.5$ V (– towards X)

$I_{R_1} = -5.88 + 1.18 = 4.7$ A (– towards X)

$V_{R_2} = 61.8 - 82.3 = 20.5$ V (– towards X)

$I_{R_2} = 1.77 - 2.35 = 0.58$ A (– towards X)

$V_{R_3} = 61.8 + 17.7 = 79.5$ V (+ towards X)

$I_{R_3} = 4.12 + 1.18 = 5.3$ A (+ towards X)

20. $R_{TH} = 40.5\ \Omega$

$V_{TH} = 70$ V

21. $V_{TH} = 5.77$ V

$R_{TH} = 36.5\ \Omega$

22. $R_N = 38.3\ \Omega$

$I_N = 0.522$ A

23. $R_X = 600\ \Omega$

ANSWERS TO STUDY GUIDE

Pages 79–92

1. d.
2. c.
3. a.
4. b.
5. a.
6. b.
7. a.
8. b.
9. c.
10. d.
11. b.

12. a.

13. Current equation: $I_1 + I_2 = I_3 + I_4$
Substitution: $0.5 \text{ A} + 0.3 \text{ A} = I_3 + 0.2 \text{ A}$
Answer: $I_3 = 0.6$ A

14. a. Currents at junction A:
Current equation: $I_1 = I_2 + I_3$
Substitution: $I_1 = 150 \text{ mA} + 250 \text{ mA}$
Answer: $I_1 = 400$ mA

  b. Currents at junction C:
Current equation: $I_3 = I_4 + I_6$
Substitution: $250 \text{ mA} = I_4 + 200 \text{ mA}$
Answer: $I_4 = 50$ mA (away from junction C)

  c. Currents at junction B:
Current equation: $I_5 = I_2 + I_4$
Substitution: $I_5 = 150 \text{ mA} + 50 \text{ mA}$
Answer: $I_5 = 200$ mA (away from junction B)

  d. Currents at junction D:
Current equation: $I_7 = I_5 + I_6$
Substitution: $I_7 = 200 \text{ mA} + 200 \text{ mA}$
Answer: $I_7 = 400$ mA (away from junction D)

15. Loop equation: $V - V_{R_3} - V_{R_2} - V_{R_1} = 0$
Substitution: $10 \text{ V} - 3 \text{ V} - 2V - V_{R_3} = 0$
Answer: $V_{R_3} = 5$ V

16. a. Loop A:
Loop equation: $-V + V_{R_5} + V_{R_2} + V_{R_1} = 0$
Substitution: $-80 \text{ V} + 50 \text{ V} + V_{R_2} + 20 \text{ V} = 0$
Answer: $V_{R_2} = 10$ V

  b. Loop B:
Equation: $V_{R_2} - V_{R_3} - V_{R_4} = 0$
Substitution: $10 \text{ V} - V_{R_3} - 2 \text{ V} = 0$
Answer: $V_{R_3} = 8$ V

17. a. $V_{R_1} = I_1 \times R_1$
$V_{R_2} = I_2 \times R_2$
$V_{R_3} = I_3 \times R_3$

  b. Loop A:
Equation: $I_A R_1 + (I_A + I_B)R_3 - V_A = 0$
Substitution: $40I_A + 20(I_A + I_B) = 60$
Simplification: $40I_A + 20I_A + 20I_B = 60$
$60I_A + 20I_B = 60$

  c. Loop B:
Equation: $I_B R_2 + (I_A + I_B)R_3 - V_B = 0$
Substitution: $20I_B + 20(I_A + I_B) = 80$
Simplification: $20I_B + 20I_A + 20I_B = 80$
$20I_A + 40I_B = 80$

  d. Combining:
Multiply equation for loop B by 3:
$60I_A + 120I_B = 240$
Subtract the equations:
$100I_B = 180$
$I_B = 1.8$ A
Substituting for I_B:
$20I_A + 40I_B = 80$
$20I_A + 40(1.8) = 80$
$I_A = 0.4$ A

  e. Currents:
$I_{R_1} = 0.4$ A
$I_{R_2} = 1.8$ A
$I_{R_3} = 2.2$ A

  f. Voltage drop across R_1:
Formula: $V_{R_1} = I_1 \times R_1$
Substitution: $V_{R_1} = 0.4 \times 40 \ \Omega$
Answer: $V_{R_1} = 16$ V

  g. Voltage drop across R_2:
Answer: $V_{R_2} = 36$ V

  h. Voltage drop across R_3:
Answer: $V_{R_3} = 44$ V

18. Part 1: Student drawing with V_B replaced with a short.

    a. Total resistance for voltage source A has R_2 in parallel with R_3. The combination is in series with R_1.

       Formula: $R_{T_A} = R_1 + R_2 \parallel R_3$

       Substitution: $R_{T_A} = 10\ \Omega + 10.5\ \Omega$

       Answer: $R_{T_A} = 20.5\ \Omega$

    b. Total current from V_A:

       Formula: $I_{T_A} = \dfrac{V_A}{R_{T_A}}$

       Substitution: $I_{T_A} = \dfrac{100\ \text{V}}{20.5\ \Omega}$

       Answer: $I_{T_A} = 4.88\ \text{A}$

    c. Voltage drop across R_1 from V_A:

       Formula: $V_{R_{1_A}} = I_{T_A} \times R_1$

       Substitution: $V_{R_{1_A}} = 4.88\ \text{A} \times 10\ \Omega$

       Answer: $V_{R_{1_A}} = 48.8\ \text{V}$

    d. Voltage across R_2 and R_3 from V_A:

       Formula: $V_{R_{2_A}} = V_A - V_{R_{1_A}}$

       Substitution: $V_{R_{2_A}} = 100\ \text{V} - 48.8\ \text{V}$

       Answer: $V_{R_{2_A}} = V_{R_{3_A}} = 51.2\ \text{V}$

    e. Current in R_2 from V_A:

       Formula: $I_{R_{2_A}} = \dfrac{V_{R_{2_A}}}{R_2}$

       Substitution: $I_{R_{2_A}} = \dfrac{51.2\ \text{V}}{35\ \Omega}$

       Answer: $I_{R_{2_A}} = 1.46\ \text{A}$

    f. Current in R_3 from V_A:

       Formula: $I_{R_{3_A}} = \dfrac{V_{R_{3_A}}}{R_3}$

       Substitution: $I_{R_{3_A}} = \dfrac{51.2\ \text{V}}{15\ \Omega}$

       Answer: $I_{R_{3_A}} = 3.41\ \text{A}$

Part 2: Student drawing with V_A replaced with a short.

    a. Total resistance for voltage source A has R_1 in parallel with R_3. The combination is in series with R_2.

       Formula: $R_{T_B} = R_2 + R_1 \parallel R_3$

       Substitution: $R_{T_B} = 35\ \Omega + 6\ \Omega$

       Answer: $R_{T_B} = 41\ \Omega$

    b. Total current from V_B:

       Formula: $I_{T_B} = \dfrac{V_B}{R_{T_B}}$

       Substitution: $I_{T_B} = \dfrac{50\ \text{V}}{41\ \Omega}$

       Answer: $I_{T_B} = 1.22\ \text{A}$

    c. Voltage drop across R_2 from V_B:

       Formula: $V_{R_{2_B}} = I_{T_B} \times R_2$

       Substitution: $V_{R_{2_B}} = 1.22\ \text{A} \times 35\ \Omega$

       Answer: $V_{R_{2_B}} = 42.7\ \text{V}$

    d. Voltage drop across R_1 and R_3 from V_B:

       Formula: $V_{R_{1_B}} = V_B - V_{R_{2_B}}$

       Substitution: $V_{R_{1_B}} = 50\ \text{V} - 42.7\ \text{V}$

       Answer: $V_{R_{1_B}} = V_{R_{3_B}} = 7.3\ \text{V}$

    e. Current in R_1 from V_B:

       Formula: $I_{R_{1_B}} = \dfrac{V_{R_{1_B}}}{R_1}$

       Substitution: $I_{R_{1_B}} = \dfrac{7.3\ \text{V}}{10\ \Omega}$

       Answer: $I_{R_{1_B}} = 0.73\ \text{A}$

    f. Current in R_3 from V_B:

       Formula: $I_{R_{3_B}} = \dfrac{V_{R_{3_B}}}{R_3}$

       Substitution: $I_{R_{3_B}} = \dfrac{7.3\ \text{V}}{15\ \Omega}$

       Answer: $I_{R_{3_B}} = 0.487\ \text{A}$

Part 3: Combining:

$$V_{R_1} = V_{R_{1_A}} - V_{R_{1_B}} = 48.8 \text{ V} - 7.3 \text{ V} = 41.5 \text{ V}$$
(– towards X)

$$V_{R_2} = V_{R_{2_B}} - V_{R_{2_A}} = 42.7 \text{ V} - 51.2 \text{ V} = -8.5 \text{ V}$$
(+ towards X)

$$V_{R_3} = V_{R_{3_A}} + V_{R_{3_B}} = 51.2 \text{ V} + 7.3 \text{ V} = 58.5 \text{ V}$$
(+ towards X)

$$I_{R_1} = I_{T_A} - I_{R_{1_B}} = 4.88 \text{ A} - 0.73 \text{ A} = 4.15 \text{ A}$$
(away from X)

$$I_{R_2} = I_{T_B} - I_{R_{2_A}} = 1.22 \text{ A} - 1.46 \text{ A} = -0.24$$
(towards X)

$$I_{R_3} = I_{R_{3_A}} + I_{R_{3_B}} = 3.41 \text{ A} + 0.487 \text{ A} = 3.90 \text{ A}$$
(towards X)

Note: to check for accuracy, use Kirchhoff's laws.

$$V_A = V_{R_1} + V_{R_3}$$
$$100 \text{ V} = 41.5 \text{ V} + 58.5 \text{ V}$$

$$V_B = V_{R_2} + V_{R_3}$$
$$50 \text{ V} = -8.5 \text{ V} + 58.5 \text{ V}$$

$$I_{R_3} = I_{R_1} + I_{R_2}$$
$$3.90 \text{ A} = 4.15 \text{ A} - 0.24 \text{ A}$$

19. a. Thevenin voltage:

Formula: $V_{TH} = \dfrac{V \times R_3}{R_1 + R_3}$

Substitution: $V_{TH} = \dfrac{25 \text{ V} \times 15 \text{ Ω}}{10 \text{ Ω} + 15 \text{ Ω}}$

Answer: $V_{TH} = 15 \text{ V}$

b. R_{TH} is found by removing R_L and looking back from the terminals with the voltage shorted. R_1 is in parallel with R_3. The combination is in series with R_2.
$R_{TH} = 31 \text{ Ω}$

c. Student sketch of Thevenin equivalent circuit.

d. Total resistance of equivalent circuit, including load:
Formula: $R_T = R_{TH} + R_L$
Substitution: $R_T = 31 \text{ Ω} + 5 \text{ Ω}$
Answer: $R_T = 36 \text{ Ω}$

e. Load current:

Formula: $I_L = \dfrac{V_{TH}}{R_T}$

Substitution: $I_L = \dfrac{15 \text{ V}}{36 \text{ Ω}}$

Answer: $I_L = 0.417 \text{ A}$

f. Load voltage:
Formula: $V_L = I_L \times R_L$
Substitution: $V_L = 0.417 \text{ A} \times 5 \text{ Ω}$
Answer: $V_L = 2.09 \text{ V}$

20. a. V_{TH} is found by replacing the load with a voltmeter. Then, find the difference between the voltage across R_3 and the voltage across R_4.
$V_{R_3} = 50 \text{ V}$
$V_{R_4} = 66.7 \text{ V}$
$V_{TH} = 16.7 \text{ V}$ (+ on terminal B)

b. R_{TH} is found by removing the load and looking back into the terminals with the voltage replaced by a short. R_1 is in parallel with R_3. R_2 is in parallel with R_4. The two combinations are in series with each other.
$R_{TH} = 38.3 \text{ Ω}$

c. Student sketch of Thevenin equivalent circuit.

d. Total resistance of equivalent circuit with load.
Formula: $R_T = R_{TH} + R_L$
Substitution: $R_T = 38.3 \text{ Ω} + 10 \text{ Ω}$
Answer: $R_T = 48.3 \text{ Ω}$

e. Load current:

Formula: $I_L = \dfrac{V_{TH}}{R_T}$

Substitution: $I_L = \dfrac{16.7 \text{ V}}{48.3 \text{ Ω}}$

Answer: $I_L = 0.346 \text{ A}$

f. Load voltage:
Formula: $V_L = I_L \times R_L$
Substitution: $V_L = 0.346 \text{ A} \times 10 \text{ Ω}$
Answer: $V_L = 3.46 \text{ V}$

21. Part 1. Norton current:
   a. Total resistance with load shorted.
   $R_T = R_1 + R_2 \parallel R_3$
   $R_T = 80 \ \Omega$

   b. I_T with load shorted.

   Formula: $I_T = \dfrac{V}{R_T}$

   Substitution: $I_T = \dfrac{120 \ V}{80 \ \Omega}$

   Answer: $I_T = 1.5 \ A$

   c. Voltage across mainline resistor.
   Formula: $V_{R_1} = I_T \times R_1$
   Substitution: $V_{R_1} = 1.5 \ A \times 60 \ \Omega$
   Answer: $V_{R_1} = 90 \ V$

   d. Voltage across parallel branches.
   $V_{R_3} = V_{R_2} = 30 \ V$

   e. Current through short replacing load.
   $I_N = 1 \ A \ \Omega$

   Part 2. Norton resistance:

   a. Norton resistance is the same as Thevenin resistance. Remove the load and look back with the voltage shorted.
   $R_N = R_1 \parallel R_3 + R_2$
   $R_N = 30 \ \Omega + 30 \ \Omega$
   $R_N = 60 \ \Omega$

   Part 3. Norton equivalent circuit:
   Student drawing of equivalent circuit.

   a. Total resistance of equivalent circuit with load:

   Formula: $R_T = R_N \parallel R_L$

   Substitution: $R_T = \dfrac{60 \ \Omega \times 20 \ \Omega}{60 \ \Omega + 20 \ \Omega}$

   Answer: $R_T = 15 \ \Omega$

   b. Circuit voltage:
   Formula: $V = I_N \times R_T$
   Substitution: $V = 1 \ A \times 15 \ \Omega$
   Answer: $V = 15 \ V$

   c. Load current:

   Formula: $I_L = \dfrac{V}{R_L}$

   Substitution: $I_L = \dfrac{15 \ V}{20 \ \Omega}$

   Answer: $I_L = 0.75 \ A$

22. Formula: $\dfrac{R_X}{R_2} = \dfrac{R_1}{R_3}$

   Substitution: $\dfrac{R_X}{1000 \ \Omega} = \dfrac{10 \ \Omega}{50 \ \Omega}$

   Answer: $R_X = 200 \ \Omega$

ANSWERS TO CHAPTER TEST IN THE INSTRUCTOR'S MANUAL

Pages 129–143
1. zero
2. zero
3. separately
4. removed, open
5. short
6. series
7. voltage, current
8. short
9. parallel

10. Current equation: $I_1 + I_2 - I_3 - I_4 = 0$
    Substitution: $1.2 \ A + 0.8 \ A - 0.6 \ A - I_4 = 0$
    Answer: $I_4 = 1.4 \ A$

11. Loop equation: $-V_{R_5} + -V_{R_2} + -V_{R_1} + V = 0$ (Loop A)
    Substitution: $-30 \ V + -V_{R_2} + -10 \ V + 60 \ V = 0$
    Answer: $V_{R_2} = 20 \ V$

12. Loop equation: $V_{R_3} + V_{R_4} - V_{R_2} = 0$ (Loop B)
    Substitution: $V_{R_3} + 5 \ V - 20 \ V = 0$
    Answer: $V_{R_3} = 15 \ V$

13.   a.   Redraw the circuit with voltage source V_B replaced with a short.

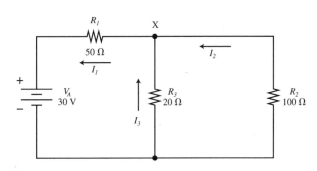

b.   Show steps to find the total resistance from V_A. Find $R_2 \parallel R_3$:

Formula: $\dfrac{1}{R_{2-3}} = \dfrac{1}{R_2} + \dfrac{1}{R_3}$

Substitution: $\dfrac{1}{R_{2-3}} = \dfrac{1}{100\ \Omega} + \dfrac{1}{20\ \Omega}$

Answer: $R_{2-3} = 16.7\ \Omega$

Combine with series R_1:
Formula: $R_{T_A} = R_1 + R_{2-3}$
Substitution: $R_{T_A} = 50\ \Omega + 16.7\ \Omega$
Answer: $R_{T_A} = 66.7\ \Omega$

c.   Total current from V_A:

Formula: $I_{T_A} = \dfrac{V_A}{R_{T_A}}$

Substitution: $I_{T_A} = \dfrac{30\ V}{66.7\ \Omega}$

Answer: $I_{T_A} = 0.45\ A$

d.   Voltage drop across R_1 from V_A:

Formula: $V_{R_{1_A}} = I_{T_A} \times R_1$
Substitution: $V_{R_{1_A}} = 0.45\ A \times 50\ \Omega$
Answer: $V_{R_{1_A}} = 22.5\ V$

e.   Voltage across R_2 and R_3 from V_A:

Formula: $V_{R_{2-3_A}} = I_{T_A} \times R_{2-3}$
Substitution: $V_{R_{2-3_A}} = 0.45\ A \times 16.7\ \Omega$
Answer: $V_{R_{2-3_A}} = 7.5\ V$

f.   Current in R_2 from V_A:

Formula: $I_{R_{2_A}} = \dfrac{V_{R_{2-3_A}}}{R_2}$

Substitution: $I_{R_{2_A}} = \dfrac{7.5\ V}{100\ \Omega}$

Answer: $I_{R_{2_A}} = 0.075\ A$

g.   Current in R_3 from V_A:

Formula: $I_{R_{3_A}} = \dfrac{V_{R_{2-3_A}}}{R_3}$

Substitution: $I_{R_{3_A}} = \dfrac{7.5\ V}{20\ \Omega}$

Answer: $I_{R_{3_A}} = 0.375\ A$

14.   a.   Redraw the circuit with voltage source V_A replaced with a short.

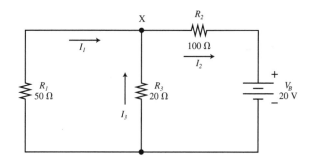

b.   Show steps to find the total resistance from V_B. Combine parallel $R_1 \parallel R_3$:

Formula: $\dfrac{1}{R_{1-3}} = \dfrac{1}{R_1} + \dfrac{1}{R_3}$

Substitution: $\dfrac{1}{R_{1-3}} = \dfrac{1}{50\ \Omega} + \dfrac{1}{20\ \Omega}$

Answer: $R_{1-3} = 14.3\ \Omega$

Combine with series R_2:
Formula: $R_{T_B} = R_{1-3} + R_2$
Substitution: $R_{T_B} = 14.3\ \Omega + 100\ \Omega$
Answer: $R_{T_B} = 114.3\ \Omega$

c. Total current from V_B:

Formula: $I_{T_B} = \dfrac{V_B}{R_{T_B}}$

Substitution: $I_{T_B} = \dfrac{20\ \text{V}}{114.3\ \Omega}$

Answer: $I_{T_B} = 0.175\ \text{A}$

d. Voltage drop across R_2 from V_B:

Formula: $V_{R_{2_B}} = I_{T_B} \times R_2$

Substitution: $V_{R_{2_B}} = 0.175\ \text{A} \times 100\ \Omega$

Answer: $V_{R_{2_B}} = 17.5\ \text{V}$

e. Voltage across R_1 and R_3 from V_B:

Formula: $V_{R_{1-3_A}} = I_{T_B} \times R_{1-3}$

Substitution: $V_{R_{1-3_A}} = 0.175\ \text{A} \times 14.3\ \Omega$

Answer: $V_{R_{1-3_A}} = 2.5\ \text{V}$

f. Current in R_1 from V_B:

Formula: $I_{R_{1_B}} = \dfrac{V_{R_{1-3_B}}}{R_1}$

Substitution: $I_{R_{1_B}} = \dfrac{2.5\ \text{V}}{50\ \Omega}$

Answer: $I_{R_{1_B}} = 0.05\ \text{A}$

g. Current in R_3 from V_B:

Formula: $I_{R_{3_B}} = \dfrac{V_{R_{1-3_A}}}{R_3}$

Substitution: $I_{R_{3_B}} = \dfrac{2.5\ \text{V}}{20\ \Omega}$

Answer: $I_{R_{3_B}} = 0.125\ \text{A}$

15. $V_{R_1} = V_{R_{1_A}} + V_{R_{1_B}}$

$V_{R_1} = -22.5\ \text{V} + 2.5\ \text{V}$

$V_{R_1} = -20\ \text{V}$

$V_{R_2} = V_{R_{2_A}} + -V_{R_{2_B}}$

$V_{R_2} = 7.5\ \text{V} + -17.5\ \text{V}$

$V_{R_2} = -10\ \text{V}$

$V_{R_3} = V_{R_{3_A}} + V_{R_{3_B}}$

$V_{R_3} = 7.5\ \text{V} + 2.5\ \text{V}$

$V_{R_3} = 10\ \text{V}$

$I_{R_1} = I_{R_{1_A}} + I_{R_{1_B}}$

$I_{R_1} = -0.45\ \text{A} + 0.05\ \text{A}$

$I_{R_1} = -0.4\ \text{A}$

$I_{R_2} = I_{R_{2_A}} + I_{R_{2_B}}$

$I_{R_2} = 0.075\ \text{A} + -0.175\ \text{A}$

$I_{R_2} = -0.1\ \text{A}$

$I_{R_3} = I_{R_{3_A}} + I_{R_{3_B}}$

$I_{R_3} = 0.375\ \text{A} + 0.125\ \text{A}$

$I_{R_3} = 0.5\ \text{A}$

16. Remove load and leave terminals open. Use the voltage divider formula to find the voltage seen by the load terminals across R_3.

Formula: $V_{R_3} = V_{TH} = V_A \times \dfrac{R_3}{R_1 + R_3}$

Substitution: $V_{TH} = 75\ \text{V} \times \dfrac{30\ \Omega}{20\ \Omega + 30\ \Omega}$

Answer: $V_{TH} = 45\ \text{V}$

17. Combine parallel $R_1 \parallel R_3$:

Formula: $\dfrac{1}{R_{1-3}} = \dfrac{1}{R_1} + \dfrac{1}{R_3}$

Substitution: $\dfrac{1}{R_{1-3}} = \dfrac{1}{20\ \Omega} + \dfrac{1}{30\ \Omega}$

Answer: $R_{1-3} = 12\ \Omega$

Combine with series R_2:

Formula: $R_{TH} = R_2 + R_{1-3}$

Substitution: $R_{TH} = 40\ \Omega + 12\ \Omega$

Answer: $R_{TH} = 52\ \Omega$

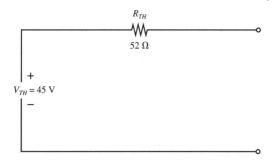

18. a. Total resistance when load is replaced by a short:
         Combine parallel $R_2 \parallel R_3$:

         Formula: $\dfrac{1}{R_{2-3}} = \dfrac{1}{R_2} + \dfrac{1}{R_3}$

         Substitution: $\dfrac{1}{R_{2-3}} = \dfrac{1}{80\ \Omega} + \dfrac{1}{80\ \Omega}$

         Answer: $R_{2-3} = 40\ \Omega$

         Combine with series R_1:
         Formula: $R_T = R_1 + R_{2-3}$
         Substitution: $R_T = 40\ \Omega + 40\ \Omega$
         Answer: $R_T = 80\ \Omega$

     b. Total current when load is replaced by a short:

         Formula: $I_T = \dfrac{V}{R_T}$

         Substitution: $I_T = \dfrac{20\ V}{80\ \Omega}$

         Answer: $I_T = 0.25\ A$

     c. Voltage drop across the mainline resistor:
         Formula: $V_{R_1} = I_T \times R_1$
         Substitution: $V_{R_1} = 0.25\ A \times 40\ \Omega$
         Answer: $V_{R_1} = 10\ V$

     d. Voltage drop across the parallel combination:
         Formula: $V_{R_{2-3}} = I_T \times R_{2-3}$
         Substitution: $V_{R_{2-3}} = 0.25\ A \times 40\ \Omega$
         Answer: $V_{R_{2-3}} = 10\ V$

     e. Find Norton's current (the current through the short replacing the load).

         Formula: $I_N = \dfrac{V_{R_{2-3}}}{R_2}$

         Substitution: $I_N = \dfrac{10\ V}{80\ \Omega}$

         Answer: $I_N = 0.125\ A$

19. Remove voltage source and replace with a short.
     Reminder: $R_N = R_{TH}$
     Combine parallel $R_1 \parallel R_3$:

     Formula: $\dfrac{1}{R_{1-3}} = \dfrac{1}{R_1} + \dfrac{1}{R_3}$

     Substitution: $\dfrac{1}{R_{1-3}} = \dfrac{1}{40\ \Omega} + \dfrac{1}{80\ \Omega}$

     Answer: $R_{1-3} = 26.7\ \Omega$

     Combine with series R_2:
     Formula: $R_N = R_2 + R_{1-3}$
     Substitution: $R_N = 80\ \Omega + 26.7\ \Omega$
     Answer: $R_N = 106.7\ \Omega$

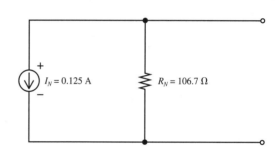

20. Formula: $\dfrac{R_X}{R_2} = \dfrac{R_1}{R_3}$

     Substitution: $\dfrac{R_X}{400\ \Omega} = \dfrac{20\ \Omega}{80\ \Omega}$

     Answer: $R_X = 100\ \Omega$

Chapter Test

9

DC Circuit Theorems

Fill in the blanks.

1. Kirchhoff's current law states the algebraic sum of the currents entering and leaving a point will equal _____.

2. Kirchhoff's voltage law states the algebraic sum of the voltages around a loop will equal _____.

3. In a circuit containing more than one voltage source, the superposition theorem states the current or voltage of individual components is the algebraic sum of the sources acting _____.

4. The voltage in a Thevenin equivalent circuit is calculated with the load _____

   and the terminals _____.

5. The Thevenin equivalent resistance is found by removing the load and the voltage. The voltage is replaced with a

   _____.

6. When drawing the Thevenin equivalent circuit, the Thevenin resistance is connected in _____

   (series, parallel) with the Thevenin voltage.

7. After the Thevenin equivalent circuit has been found, only the values of the load _____ and

   the _____ will change from one load to another.

8. The Norton current is found by removing the load and replacing it with a _____.

9. The Norton equivalent resistance is drawn by connecting it in _____ (series, parallel) with the

   Norton current source.

With each problem, write the equation, substitution, and answer.

10. Use Kirchhoff's current law to find current I_4 in the following figure.

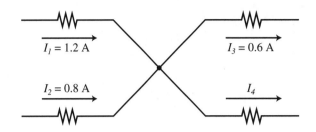

Current equation: _____

Substitution: _____

Answer: _____

11. Use Kirchhoff's voltage law to find V_{R_2} in the following figure.

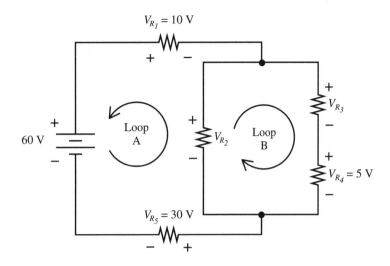

Loop equation: _____

Substitution: _____

Answer: _____

12. Use Kirchhoff's voltage law to find V_{R_3} in the figure shown in question 11.

Loop equation: _____

Substitution: _____

Answer: _____

For questions 13, 14, and 15, use the superposition theorem to find the current and voltage of each resistor in the following figure.

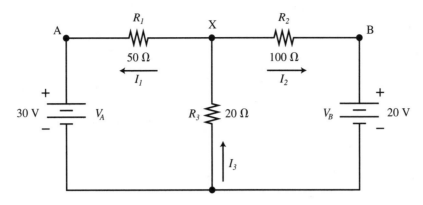

13. a. Redraw the circuit with voltage source V_B replaced with a short.

   b. Show steps to find the total resistance from V_A.

_____

_____

_____

_____

_____

_____

c. Total current from V_A:

Formula: _____

Substitution: _____

Answer: _____

d. Voltage drop across R_1 from V_A:

Formula: _____

Substitution: _____

Answer: _____

e. Voltage across R_2 and R_3 from V_A:

Formula: _____

Substitution: _____

Answer: _____

f. Current in R_2 from V_A:

Formula: _____

Substitution: _____

Answer: _____

Name: ______________________

g. Current in R_3 from V_A:

Formula: ______________________

Substitution: ______________________

Answer: ______________________

14. a. Redraw the circuit with voltage source V_A replaced with a short.

b. Show steps to find the total resistance from V_B.

____

____

____

____

____

____

c. Total current from V_B:

Formula: _____

Substitution: _____

Answer: _____

d. Voltage drop across R_2 from V_B:

Formula: _____

Substitution: _____

Answer: _____

e. Voltage across R_1 and R_3 from V_B:

Formula: _____

Substitution: _____

Answer: _____

f. Current in R_1 from V_B:

Formula: _____

Substitution: _____

Answer: _____

g. Current in R_3 from V_B:

Formula: _____

Substitution: _____

Answer: _____

15. Superimpose voltages and currents for the circuit of questions 13 and 14. Find the requested values.

$V_{R_1} =$ _____

$V_{R_2} =$ _____

$V_{R_3} =$ _____

$I_{R_1} =$ _____

$I_{R_2} =$ _____

$I_{R_3} =$ _____

For questions 16 and 17, find the Thevenin equivalent circuit of the following figure.

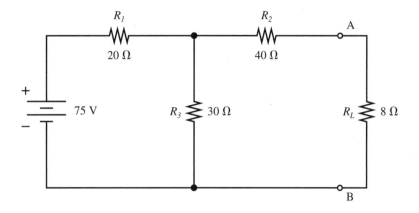

16. Show the steps to find the Thevenin voltage. _____

_____

_____

_____

_____

17. Show the steps to find the Thevenin resistance. _____

_____

_____

_____

_____

_____

_____

_____

Draw the Thevenin equivalent circuit.

For questions 18 and 19, find the Norton equivalent circuit of the following figure.

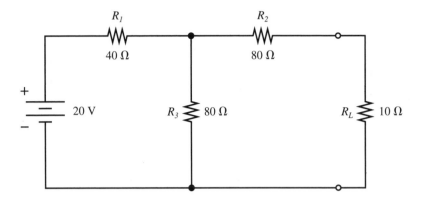

18. Show the steps to find the Norton current.

    a. Total resistance when load is replaced by a short:

    _____

    _____

    _____

    _____

    _____

    _____

    b. Total current when load is replaced by a short:

    Formula: _____

    Substitution: _____

    Answer: _____

c. Voltage drop across the mainline resistor:

Formula: _____

Substitution: _____

Answer: _____

d. Voltage drop across the parallel combination:

Formula: _____

Substitution: _____

Answer: _____

e. Find Norton's current (the current through the short replacing the load).

Formula: _____

Substitution: _____

Answer: _____

19. Show the steps to find the Norton resistance.

_____

_____

_____

_____

_____

_____

Draw the Norton equivalent circuit.

20. Find the value of R_X needed to balance the Wheatstone bridge.

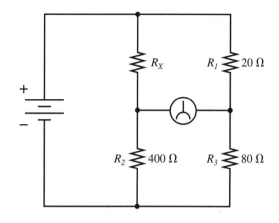

Formula: _____

Substitution: _____

Answer: _____

Producing DC Voltages

OBJECTIVES

After studying this chapter, students should be able to:
- Relate several individuals to important inventions and developments in producing electricity.
- Describe the characteristics and behavior of static electricity.
- Identify examples of the piezoelectric effect.
- Give examples where the effects of heat can be used.
- Describe applications of light-sensitive devices.
- Recognize devices using magnetism to develop electrical signals.
- Describe the differences between different batteries.
- Calculate the results of voltage sources connected in series and parallel.
- Predict the effect of a voltage source's internal resistance.
- Identify the maximum transfer of power.

INSTRUCTIONAL MATERIALS

Text: Pages 283–310
Test Your Knowledge Questions, Pages 308–309
Study Guide: Pages 93–96
Laboratory Manual: Pages 95–112

ANSWERS TO TEXTBOOK

Test Your Knowledge, Pages 308–309
1. a. Thomas Edison: Developed an alkaline secondary cell.
   b. Luigi Galvani: Found electricity activates animal nerves and muscles.
   c. Gaston Planté: Invented the lead-acid wet-cell battery.
   d. Robert Van de Graaff: Developed an electrostatic machine.
   e. Benjamin Franklin: Proved lightning is electricity, invented the lightning rod, and developed a theory to explain positive and negative electrical charges.
   f. Georges Lechanché: Invented the primary cell.
   g. Thomas Seebeck: Observed current flows when dissimilar metals are connected and heated.
   h. Alessandro Volta: Developed the earliest battery.
2. A static charge is developed in insulators, such as glass or rubber, by gathering electrons in a material that allows little movement (often through friction).
3. Insulators.
4. a. Discharge through a wire.
   b. Discharge through touching.
   c. Discharge through arcing.
5. Piezoelectric effect.
6. Sound waves strike a diaphragm, which puts pressure on the crystal. The crystal produces a voltage.
7. Radio frequency crystals have the ability to produce accurate frequencies.
8. a. Thermistor varies resistance with heat.
   b. Thermocouple produces a voltage with heat.
   c. Thermostat is a heat-activated switch.
9. a. The thermistor is a resistor that is sensitive to changes in heat. A typical application is a thermometer.
   b. A thermocouple is two different pieces of metal joined at one end. When heat is applied a voltage is produced. A typical application is as a heat sensor, such as an electronically controlled burner.
   c. The thermostat is a heat-activated switch. A typical application is to turn a furnace on and off when the room temperature reaches the desired value.

10. Photosensitive.

11. Optoelectric.

12. A photocell responds to the light and dark areas on the coded bar using the reflected light.

13. The magnetic tape induces a voltage in the magnetic pickup as it passes by.

14. Student lists will vary.

15. Electrolyte.

16. Some ions are positively charged, some are negatively charged. When a path is provided the ions will flow.

17. The wet cell uses a liquid as an electrolyte and is easily recharged. The dry cell uses a powder as an electrolyte and is not rechargeable.

18. To recharge a battery, an outside voltage forces current through the electrolyte in the direction opposite to which the current flows during discharge.

19. 9 volts.

20. Less than 10 hours.

21. Connect in parallel to maintain the same voltage and increase the ampere-hour rating. The new estimated time would be less than 20 hours.

22. Internal resistance is measured in the forward direction, when the battery is being discharged. Charging resistance is the resistance in the reverse direction needed to overcome the resistance of the electrolyte.

23. Internal resistance increases with a weaker battery.

24. As current flows to the load voltage drops across the internal resistance. Ohm's law states that a larger current produces more voltage drop. The voltage across the internal resistance subtracts from the load voltage.

25. Maximum transfer of power happens when the load resistance equals the internal resistance.

ANSWERS TO STUDY GUIDE

Pages 93–96

1. k.
2. d.
3. b.
4. l.
5. h.
6. m.
7. n.
8. o.
9. g.
10. f.

11. c.
12. e.
13. j.
14. i.
15. a.
16. e.
17. c.
18. a.
19. g.
20. b.
21. f.
22. d.
23. h.
24. c.
25. d.
26. a.
27. a.
28. a.
29. b.
30. b.
31. b.
32. d.
33. c.
34. a.
35. a.
36. d.
37. b.
38. c.
39. c.
40. d.
41. a.
42. a.

ANSWERS TO CHAPTER TEST IN THE INSTRUCTOR'S MANUAL

Pages 149–150

1. Primary cell: A single cell of a battery that cannot be recharged.

2. Secondary cell: A single cell of a battery that can be recharged easily.

3. Electrolyte: A substance that produces ions when it conducts electricity.

4. Thermistor: A resistor that varies its resistance with changes in temperature.

5. Optoelectric: Electronic components that use light to operate.

6. Thermocouple: Two different metals connected together that produce electricity when heated.

7. Photosensitive: Materials that respond to light strikes by either producing a voltage or changing resistance.

8. Thermostat: A heat-activated switch.

9. Piezoelectric effect: Voltage produced in crystals when a mechanical pressure is applied.

10. Static electricity: A buildup or shortage of electrons producing a difference in potential until a discharge path is provided.

11. Thomas Edison: Developed a secondary cell, which can be recharged.

12. Benjamin Franklin: Proved that lightning is an electrical phenomenon; invented the lightning rod; developed the *one-fluid theory*.

13. Thomas Seebeck: Discovered that when two dissimilar metals are connected and heated they produce a voltage.

14. Luigi Galvani: Found that electricity activates animal nerves and muscles.

15. Georges Leclanché: Invented the primary cell.

16. hole

17. static

18. piezoelectric

19. photosensitive

20. optoelectric

21. reflected light

22. voltage

23. six

24. five

25. load, internal

Chapter Test

10

Producing DC Voltages

Define these technical terms.

1. Primary cell: _____

   _____

2. Secondary cell: _____

   _____

3. Electrolyte: _____

   _____

4. Thermistor: _____

   _____

5. Optoelectric: _____

   _____

6. Thermocouple: _____

   _____

7. Photosensitive: _____

   _____

8. Thermostat: _____

   _____

9. Piezoelectric effect: _____

   _____

10. Static electricity: _____

    _____

Make a brief statement of the contribution made by the following individuals, as related to this chapter.

11. Thomas Edison: _____

_____

12. Benjamin Franklin: _____

_____

13. Thomas Seebeck: _____

_____

14. Luigi Galvani: _____

_____

15. Georges Leclanché:_____

_____

Fill in the blanks.

16. A(n) _____ is an atom seeking an electron.

17. A buildup of electrons in a nonconductive material will produce a(n) _____ charge.

18. The _____ effect is pressure applied to a crystal to produce a voltage.

19. A(n) _____ material responds to light.

20. A(n) _____ device is an electronic component that uses light-sensitive materials to operate.

21. A code reader is able to read the coded bars by sensing _____.

22. The information stored on magnetic tape is changed to electrical signals by inducing _____

    in the coils of the head.

23. A portable stereo system that requires four D-size batteries connected in series, operates using _____

    volts.

24. A 6 volt emergency light requires 2 amps to operate properly. It is connected to a rechargeable 6 volt battery rated

    for 10 ampere-hours. The estimated time of usefulness for the light in an emergency is _____

    hours.

25. A voltage source is able to achieve the maximum transfer of power when _____ resis-

    tance equals _____ resistance.

Magnetic Principles and Devices

OBJECTIVES

After studying this chapter, students should be able to:
- Relate the names of selected individuals to magnetic principles they helped to develop.
- Define technical terms used with magnetism.
- Describe how certain materials are affected by magnetism.
- Explain how electricity produces magnetism.
- Identify units of measure related to magnetism.
- Perform calculations using formulas related to magnetism.
- Explain the operation of basic electromagnetic devices and give their applications.
- Interpret the ratings of electromagnetic devices.

INSTRUCTIONAL MATERIALS

Text: Pages 311–346
      Test Your Knowledge Questions, Pages 344–345
Study Guide: Pages 97–102
Laboratory Manual: Pages 113–117

ANSWERS TO TEXTBOOK

Test Your Knowledge, Pages 344–345
1. Lodestone.
2. a. ampere-turns: Unit of measure of magnetomotive force, mks system.
   b. gilbert per maxwell: Unit of measure of reluctance, cgs system.
   c. ampere-turns per meter: Unit of measure of magnetic field intensity, mks system.
   d. maxwell: Unit of measure of magnetic flux, cgs system.
   e. ampere-turns per weber: Unit of measure of reluctance, mks system.
   f. henry per meter: Unit of measure of permeability, mks system.
   g. dyne: Unit of measure of force, cgs system.
   h. oersted: Unit of measure of magnetic field intensity, cgs system.
   i. gauss: Unit measure of magnetomotive force in the cgs system.
   j. tesla: Unit of measure of flux density, mks system.
   k. gauss per oersted: Unit of measure of permeability, cgs system.
   l. unit pole: Measure of the strength of a magnetic pole, cgs system.
   m. gilbert: Unit of measure of magnetomotive force, cgs system.
   n. weber: Unit of measure of magnetic flux, mks system.
3. a. Michael Faraday: Plotted the magnetic field around a conductor.
   b. Karl Gauss: Unit of measure of flux density named after him.
   c. William Gilbert: Unit of measure of magnetomotive force named after him.
   d. James Maxwell: Experimented with magnetic fields. Unit of measure of magnetic flux named in his honor.
   e. Hans Christian Oersted: Discovered existence of a magnetic field around an electrical conductor. Unit of measure of magnetizing force named in his honor.
   f. Nikola Tesla: Experimented with electricity and magnetism. Unit of measure of flux density named after him.
   g. Wilhelm Weber: Unit of measure of magnetic field strength named after him.
4. A magnet attracts ferromagnetic materials and ferrites.
5. Likes poles repel and unlike poles attract.
6. Current flowing through a wire creates an electromagnet.

7. a. More current produces a stronger magnet.
    b. Adding an iron core increases the strength.
    c. Increasing the number of turns of wire increases the strength of the magnet.
8. With the fingers of the left hand curled around a wire in the direction of the magnetic lines of flux, the thumb points in the direction of current flow.
9. With the fingers wrapped around the coil in the direction of current flow, the thumb points at the north pole.
10. $f = 104$ dynes (attracting)
11. $B = 0.18$ teslas
12. $B = 6250$ gausses
13. a. mks: mmf = 7500 ampere-turns
     b. cgs: mmf = 5955 gilberts
14. $H = 12,500$ ampere-turns per meter
15. $H = 500$ oersteds
16. $B = 0.125$ teslas
17. $\mathcal{R} = 320,000$ ampere-turns per weber
18. mmf = 200 gilberts
19. Frequency response.
     Maximum wattage.
     Ohmic value of the voice coil.
     Weight of the magnet.
20. The design of the filter is determined by what frequencies each speaker is designed to reproduce.
21. No. The resistance of a speaker is an ac resistance value, called impedance. Ohmmeters read dc resistance, which is the resistance of the coil of wire.
22. The core.

ANSWERS TO STUDY GUIDE

Pages 97–102

1. h.
2. i.
3. k.
4. t.
5. p.
6. n.
7. m.
8. f.
9. g.
10. l.
11. j.
12. o.
13. e.
14. d.
15. q.
16. b.
17. v.
18. w.
19. r.
20. s.
21. y.
22. x.
23. c.
24. z.
25. u.
26. a.
27. a.
28. g.
29. i.
30. n.
31. k.
32. b.
33. d.
34. m.
35. l.
36. e.
37. j.
38. f.
39. c.
40. h.
41. Refer to textbook figure 11-19.
42. c.
43. g.
44. a.
45. e.
46. b.
47. f.
48. d.
49. a.
50. c.
51. c.
52. Point the thumb at the north pole. Fingers are wrapped around the coil in the direction of current flow.
53. With the fingers wrapped around the coil in the direction of current flow, the thumb points at the north pole.
54. a. Frequency response.
     b. Maximum wattage.
     c. Ohmic value of the voice coil.
     d. Weight of the magnet.
55. a. Normal human ear (16 Hz to 20,000 Hz).
     b. Voice (80 Hz to 1600 Hz).
     c. Musical instruments (30 Hz to 16,000 Hz).

56. b.
57. b.

58. Formula: $f = \dfrac{m_1 \times m_2}{d^2}$

    Substitution: $f = \dfrac{90 \times 65}{(15 \text{ cm})^2}$

    Answer: $f = 26$ dynes (attracting)

59. Formula: $B = \dfrac{\Phi}{A}$

    Substitution: $B = \dfrac{750 \text{ μWb}}{0.003 \text{ m}^2}$

    Answer: $B = 0.25$ T

60. Formula: $B = \dfrac{\Phi}{A}$

    Substitution: $B = \dfrac{750{,}000 \text{ Mx}}{4 \text{ cm} \times 30 \text{ cm}}$

    Answer: $B = 6250$ G

61. Formula: mmf $= N \times I$ (same formula for both systems)
    Substitution: mmf $= 2500$ turns $\times 3$ A
    Answer: mmf $= 7500$ At (mks system)
    Conversion: Change to cgs by multiplying ampere-turns by 1.26.
    Answer: mmf $= 9450$ Gb

62. Formula: $H = \dfrac{\text{mmf}}{l}$

    Substitution: $H = \dfrac{1500 \text{ At}}{0.03 \text{ m}}$

    Answer: $H = 50{,}000$ At/m

63. Formula: $H = \dfrac{Q}{A}$

    Substitution: $H = \dfrac{30{,}000 \text{ lines}}{8 \text{ cm} \times 10 \text{ cm}}$

    Answer: $H = 375$ Oe

64. Formula: $B = \mu \times H$
    Substitution: $B = 250 \text{ μH/m} \times 800 \text{ At/m}$
    Answer: $B = 0.2$ T

65. Formula: $\mathcal{R} = \dfrac{\text{mmf}}{\Phi}$

    Substitution: $\mathcal{R} = \dfrac{1000 \text{ At}}{400 \text{ μWb}}$

    Answer: $\mathcal{R} = 2{,}500{,}000$ At/Wb

66. Formula: mmf $= \mathcal{R} \times \Phi$
    Substitution: mmf $= 0.008$ Gb/Mx $\times 60{,}000$ Mx
    Answer: mmf $= 480$ Gb

ANSWERS TO CHAPTER TEST IN THE INSTRUCTOR'S MANUAL

Pages 155–158

1. Diamagnetic: Nonmagnetic materials. Materials are actually slightly repelled by a magnetic field.
2. Paramagnetic: Nonmagnetic materials. Materials are actually slightly attracted to a magnetic field.
3. Ferromagnetic: Materials that are strongly attracted by a magnetic field and are also good conductors of electricity.
4. Ferrites: Materials that are strongly attracted by a magnetic field but are poor conductors of electricity.
5. Permeance: The ability of a material to carry magnetic lines of force.
6. Permeability: The measure of the ease with which magnetic lines of force can flow through a material.
7. Lodestone: A material with magnetic properties in its natural state.
8. Reluctance: The opposition to the flow of magnetic flux.
9. Reluctivity: The specific reluctance of a material.
10. Retentivity: The ability of a material to retain a magnetic field after the magnetizing field has been removed.
11. ferromagnetic
12. repel, attract
13. c.
14. b.

15. Formula: $B = \dfrac{\Phi}{A}$

Substitution: $B = \dfrac{2400\ \mu Wb}{0.006\ m^2}$

Answer: $B = 0.4\ T$

16. Formula: $mmf = N \times I$
Substitution: $mmf = 12{,}000\ turns \times 3\ A$
Answer: $mmf = 36{,}000\ At$

17. Formula: $H = \dfrac{mmf}{l}$

Substitution: $H = \dfrac{8000\ At}{0.050\ m}$

Answer: $H = 160{,}000\ At/m$

18. Formula: $\mathcal{R} = \dfrac{mmf}{\Phi}$

Substitution: $\mathcal{R} = \dfrac{10{,}000\ At}{500\ \mu Wb}$

Answer: $\mathcal{R} = 20{,}000{,}000\ At/Wb$

19. Formula: $mmf = \mathcal{R} \times \Phi$
Substitution: $mmf = 0.0125\ Gb/Mx \times 150{,}000\ Mx$
Answer: $mmf = 1875\ Gb$

20. In the chart below, the underlined answers are those given in the test.

MAGNETIC QUANTITY		CGS SYSTEM		MKS SYSTEM	
Name	Symbol	Name	Abbreviation	Name	Abbreviation
force	f	dyne	dyne	xxxxxxxx	xxxxx
flux	Φ	maxwell	Mx	webers	Wb
flux density	B	gauss	G	tesla	T
Magnetomotive force	mmf	gilbert	Gb	ampere-turn	At
field intensity	H	oersted	Oe	amp-turn per meter	At/m
Permeability	μ	gauss per oersted	G/Oe	henry per meter	H/m
Reluctance	$\mathcal{R}$	gilbert per maxwell	Gb/Mx	amp-turns per weber	At/Wb

Chapter Test

11

Name: _____

Date: _____

Class: _____

Magnetic Principles and Devices

Define these technical terms.

1. Diamagnetic: _____

2. Paramagnetic: _____

3. Ferromagnetic: _____

4. Ferrites: _____

5. Permeance: _____

6. Permeability: _____

7. Lodestone: _____

8. Reluctance: _____

9. Reluctivity: _____

10. Retentivity: _____

Fill in the blanks.

11. A magnet has the ability to attract _____ materials.

12. When two magnets are brought close together, what is the effect on the magnets and their fields? Like poles

    _____. Unlike poles _____.

Select the best answer.

_____ 13. Which of these items will *not* change the strength of an electromagnet?
   a. The number of turns.
   b. The amount of current.
   c. Using it as a relay.
   d. Using a core.

_____ 14. If an ohmmeter is used to test the resistance of a speaker, it will read:
  a. the rated value of the coil.
  b. the dc resistance of the wire in the coil.
  c. the ac resistance of the coil.
  d. infinity.

15. Calculate the flux density (mks system) of a magnet with 2400 microwebers in an area of 0.006 square meters.

Formula: _____

Substitution: _____

Answer: _____

16. Calculate the magnetomotive force (mks system) of a solenoid with 12,000 turns of wire and a current of 3 amps.

Formula: _____

Substitution: _____

Answer: _____

17. Find the field intensity (mks system) of an electromagnet with mmf of 8000 ampere-turns and a core length of 0.050 meters.

Formula: _____

Substitution: _____

Answer: _____

18. What is the reluctance (mks system) of a relay with a mmf of 10,000 ampere-turns and a total flux of 500 microwebers?

Formula: _____

Substitution: _____

Answer: _____

19. Calculate the magnetomotive force (cgs system) required to produce a flux density of 150,000 maxwells in a material with a reluctance of 0.0125 gilberts per maxwell.

Formula: _____

Substitution: _____

Answer: _____

20. Complete the chart showing the units of measure and symbols.

MAGNETIC QUANTITY		CGS SYSTEM		MKS SYSTEM	
Name	Symbol	Name	Abbreviation	Name	Abbreviation
force				XXXXXXXX	XXXXX
	Φ				
		gauss			
				ampere-turn	
field intensity					
	μ				
			Gb/Mx		

Chapter 12

Inductance

OBJECTIVES

After studying this chapter, students should be able to:

- Explain how a voltage is induced when a conductor is moved through a magnetic field.
- Describe how a counterelectromotive force is developed.
- Describe how a counterelectromotive force opposes a change in the induced magnetic field.
- Analyze how inductance opposes a changing current.
- Estimate the effects on a circuit with a very high resistance discharge path.
- Identify the construction characteristics of an inductor.
- Recognize the ratings of an inductor.
- Calculate the resultant when inductors are connected in series or parallel.
- Calculate the effects of mutual inductance.

INSTRUCTIONAL MATERIALS

Text: Pages 347–364
Test Your Knowledge Questions, Pages 363–364
Study Guide: Pages 103–105
Laboratory Manual: Pages 119–123

ANSWER TO TEXTBOOK

Test Your Knowledge, Pages 363–364

1. Joseph Henry.
2. a. Moving the conductor past the magnetic field.
   b. Moving the magnetic field past the conductor.
   c. Changing the strength of the magnetic field.
3. Cemf is developed by inducing a voltage in a conductor. It gets stronger with faster movement. Its polarity is opposite the applied voltage.

4. Symbol for inductance is L. Unit of measure is henry, symbol H.
5. Refer to figures 12-3 to 12-9 and the associated text.
6. A larger discharge resistance builds a larger cemf. Very high resistance results in cemf high enough to create an arc.
7. a. Number of turns. Higher number, more inductance.
   b. Core material. Better ferrous material, more inductance.
   c. Spacing between turns. Tighter, more inductance.
   d. Wire size. Larger size, more inductance.
   e. Shape of coil. Smaller gap between pole, more inductance.
   f. Number of layers. More layers, more inductance.
   g. Coil diameter. Smaller diameter, more inductance.
   h. Type of windings. A right angle between layers of windings, more inductance.
8. a. Inductance value, in henrys.
   b. DC resistance, in ohms.
   c. Maximum current, in amps.
   d. Quality (Q), no units (it is a ratio).
9. $L_T = 700\ \mu H$
10. $L_T = 71.4\ \mu H$
11. $L_T = 500\ \mu H$
12. $L_T = 90\ mH$

ANSWERS TO STUDY GUIDE

Pages 103–105

1. e.
2. c.
3. d.
4. a.
5. b.

6. a. Moving the conductor past the magnetic field.
    b. Moving the magnetic field past the conductor.
    c. Changing the strength of the magnetic field.
7. magnitude, rate
8. direction, oppose
9. magnetic, larger
10. L, henry, H
11. b.
12. a. more turns
    b. soft iron
    c. tight
    d. larger
    e. toroid
    f. many
    g. small
    h. crisscrossed
13. a. Inductance value, in henrys.
    b. DC resistance, in ohms.
    c. Maximum current, in amps.
    d. Quality (Q), no units.

14. Formula: $L_T = L_1 + L_2 + L_3 + ...L_N$
    Substitution: $L_T = 50$ mH $+ 25$ mH $+ 125$ mH
    Answer: $L_T = 200$ mH

15. Formula: $\dfrac{1}{L_T} = \dfrac{1}{L_1} + \dfrac{1}{L_2} + \dfrac{1}{L_3} + ...\dfrac{1}{L_N}$

    Substitution:
    $$\frac{1}{L_T} = \frac{1}{175 \text{ mH}} + \frac{1}{425 \text{ mH}} + \frac{1}{300 \text{ mH}}$$

    Answer: $L_T = 87.7$ mH

16. Formula: $L_T = L_1 + L_2 + 2L_M$
    Substitution: $L_T = 35$ mH $+ 25$ mH $+ (2 \times 10$ mH$)$
    Answer: $L_T = 80$ mH

17. Formula: $L_T = L_1 + L_2 - 2L_M$
    Substitution: $L_T = 50$ µH $+ 150$ µH $- (2 \times 75$ µH$)$
    Answer: $L_T = 50$ µH

ANSWERS TO CHAPTER TEST IN THE INSTRUCTOR'S MANUAL

Pages 161–163

1. Counterelectromotive force: The property of an inductor to oppose any change in the instantaneous building of the magnetic field.

2. Inductance: Converts the property of a circuit to oppose a change in current due to a counterelectromotive force.
3. Mutual inductance: The effect of a magnetic field of one inductor crossing the turns of a different inductor.
4. Phasing dots: Dots on a schematic symbol used to indicate the direction in which a coil is wound.
5. Self inductance: The property of a conductor to induce voltage within itself.
6. Induced voltage: Voltage produced in a conductor as the result of passing through a magnetic field.
7. magnetic
8. opposite
9. L, henry, H
10. voltage
11. a. more turns
    b. iron
    c. tight
    d. toroid
    e. small

12. Formula: $L_T = L_1 + L_2 + L_3$
    Substitution:
    $\quad L_T = 600$ mH $+ 475$ mH $+ 1025$ mH
    Answer: $L_T = 2100$ mH

13. Formula: $L_T = \dfrac{1}{L_1} + \dfrac{1}{L_2} + \dfrac{1}{L_3}$

    Substitution:
    $$L_T = \frac{1}{400 \text{ mH}} + \frac{1}{800 \text{ mH}} + \frac{1}{600 \text{ mH}}$$

    Answer: $L_T = 185$ mH

14. Formula: $L_T = L_1 + L_2 + 2L_M$
    Substitution: $L_T = 100$ mH $+ 120$ mH $+ 2(10$ mH$)$
    Answer: $L_T = 240$ mH

15. Formula: $L_T = L_1 + L_2 - 2L_M$
    Substitution: $L_T = 350$ mH $+ 600$ mH $- 2(250$ µH$)$
    Answer: $L_T = 949.5$ mH

Chapter Test

12

Inductance

Name: _____

Date: _____

Class: _____

Define the following terms.

1. Counterelectromotive force: _____

2. Inductance: _____

3. Mutual inductance: _____

4. Phasing dots: _____

5. Self inductance: _____

6. Induced voltage: _____

Fill in the blanks.

7. Counter electromotive force opposes any change in the instantaneous building of a(n) _____ field.

8. The polarity of the cemf is _____ of the direction of the applied voltage.

9. The letter symbol of inductance is _____.

   The unit of measure of inductance is _____.

   The abbreviation for the unit of measure of inductance is _____.

10. An inductive circuit with a very high resistance discharge path will develop a large discharge

    _____ (voltage, current).

11. Circle the construction characteristics of an inductor that will increase the inductor's value.

    a. Number of turns: (more turns or less turns).

    b. Core material: (iron or air).

    c. Spacing between turns: (tight or loose).

    d. Shape of coil: (toroid or bar).

    e. Diameter: (large or small).

With each problem, write the formula, substitution, and answer.

12. Calculate the resultant of three inductors connected in series with the following values: 600 mH, 475 mH, and 1025 mH.

Formula: _____

Substitution: _____

Answer: _____

13. Calculate the resultant of three inductors connected in parallel with the following values: 400 mH, 800 mH, and 600 mH.

Formula: _____

Substitution: _____

Answer: _____

14. Calculate the resultant of two inductors connected in series with the values of 100 mH and 120 mH with 10 mH aiding mutual inductance.

Formula: _____

Substitution: _____

Answer: _____

15. Calculate the resultant of two inductors connected in series with the values of 350 mH and 600 mH with 250 μH opposing mutual inductance.

Formula: _____

Substitution: _____

Answer: _____

Chapter 13

Capacitance

OBJECTIVES

After studying this chapter, students should be able to:
- Define capacitance.
- Describe the construction of a capacitor.
- Explain how a capacitor works.
- Identify the factors affecting capacitance.
- Relate capacitor ratings to a catalog listing.
- List types of capacitors.
- Calculate the resultant of capacitors connected in series and parallel.
- Calculate the voltages of a capacitive voltage divider.
- Test a capacitor with an ohmmeter.

INSTRUCTIONAL MATERIALS

Text: Pages 365–384
Test Your Knowledge Questions, Page 383
Study Guide: Pages 107–110
Laboratory Manual: Pages 125–129

ANSWERS TO TEXTBOOK

Test Your Knowledge, Page 383
1. potential
2. Plate.
3. Dielectric.
4. It measures how many times better a dielectric is than a vacuum.
5. A capacitor charges by electrons flowing from the negative side of the battery to the negative side of the capacitor. For each electron gathered on the negative side, an equal number of electrons leave the positive side of the capacitor and return to the battery. When the capacitor voltage equals the supply voltage, current stops flowing.
6. Electrons leave the negative side of the capacitor and return to the positive side until there is no longer any difference in potential.
7. During charge, the resistor voltage starts with the full applied voltage. As the capacitor voltage rises, the resistor voltage decreases. Resistor voltage plus capacitor voltage equal the applied voltage during charge.
8. During discharge, the resistor voltage starts with the full capacitor voltage. As the capacitor voltage drops, the resistor voltage will follow. Resistor voltage equals the capacitor voltage during discharge.
9. a. Plate surface area. More area, more capacitance.
   b. Distance between plates. Closer plates, more capacitance.
   c. Dielectric material. Higher value, more capacitance.
10. a. Dielectric strength, measured in voltage.
    b. Amount of capacitance, measured in farads.
11. General, electrolytic, and variable.
12. $C_T = 350\ \mu F$
13. $C_T = 75\ \mu F$
14. $V_{C_1} = 9$ volts (across 100 μF)
    $V_{C_2} = 3$ volts (across 300 μF)
15. Place the ohmmeter across the two leads. The meter should indicate low resistance at first, then the measured resistance will increase to near infinity as the capacitor charges.

ANSWERS TO STUDY GUIDE

Pages 107–110
1. a.
2. d.
3. b.

4. c.

5. b.

6. d.

7. a.

8. During charge, the resistor voltage starts with the full applied voltage. As the capacitor voltage rises, the resistor voltage decreases. Resistor voltage plus capacitor voltage equals the applied voltage during charge.

9. During discharge, the resistor voltage starts with the full capacitor voltage. As the capacitor voltage drops, the resistor voltage will follow. Resistor voltage equals the capacitor voltage during discharge.

10. b.

11. d.

12. a. Plate surface area—more area, more capacitance.

    b. Distance between plates—closer plates, more capacitance.

    c. Dielectric material—more efficient, more capacitance.

13. a. Dielectric strength, measured in voltage.

    b. Amount of capacitance, measured in farads.

14. Polarized, nonpolarized, and variable.

15. Formula: $C_T = C_1 + C_2 + C_3 + ... C_N$

    Substitution: $C_T = 150\ \mu F + 200\ \mu F$

    Answer: $C_T = 350\ \mu F$

16. Formula: $\dfrac{1}{C_T} = \dfrac{1}{C_1} + \dfrac{1}{C_2} + \dfrac{1}{C_3} + ... \dfrac{1}{C_N}$

    Substitution: $\dfrac{1}{C_T} = \dfrac{1}{100\ \mu F} + \dfrac{1}{300\ \mu F}$

    Answer: $C_T = 75\ \mu F$

17. Formula: $V_{C_1} = V_A \times \dfrac{C_2}{C_1 + C_2}$

    Substitution: $V_{C_1} = 12\ V \times \dfrac{300\ \mu F}{100\ \mu F + 300\ \mu F}$

    Answer: $V_{C_1} = 9$ volts (across $100\ \mu F$)

    Answer: $V_{C_2} = 3$ volts (across $300\ \mu F$)

18. Place the ohmmeter across the two leads. For a good capacitor, the meter will first indicate a low resistance. The measured resistance will increase to near infinity as the capacitor charges.

19. a.

20. c.

21. d.

22. b.

ANSWERS TO CHAPTER TEST IN THE INSTRUCTOR'S MANUAL

Pages 169–172

1. Capacitance: Ability to store an electric charge. Capacitance also opposes a change in voltage.

2. Dielectric: An electrical insulator between the plates of a capacitor.

3. Electrostatic field: The attraction between negative and positive voltages.

4. Plate: Electrical conductors forming the negative and positive surfaces of a capacitor for gathering or releasing electrons.

5.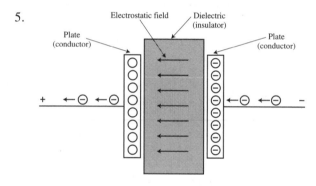

6. plates

7. dielectric

8. voltage

9. As the capacitor charges toward the applied voltage, the resistor will drop the difference in voltage between the capacitor and applied voltage.

10. As the capacitor discharges, the resistor will equal the capacitor voltage.

11. Plate surface area.
    Distance between plates.
    Dielectric material.

12. General.
    Electrolytic.
    Variable.

13. Formula: $C_T = C_1 + C_2$
    Substitution: $C_T = 0.45\ \mu F + 0.50\ \mu F$
    Answer: $C_T = 0.95\ \mu F$

14. Formula: $C_T = \dfrac{1}{C_1} + \dfrac{1}{C_2}$

    Substitution: $C_T = \dfrac{1}{400\ \mu F} + \dfrac{1}{600\ \mu F}$

    Answer: $C_T = 240\ \mu F$

15. Formula: $V_{C_1} = \dfrac{C_2}{C_1 + C_2} \times V_A$

    Substitution: $V_{C_1} = \dfrac{1400\ \mu F}{1200\ \mu F + 1400\ \mu F} \times 10\ V$

    $V_{C_1} = 5.38\ V$

    $V_{C_2} = 10\ V - 5.38\ V = 4.62\ V$

16. Short the leads of the capacitor together to remove any charge. Connect the ohmmeter across the leads of the capacitor. Observe how the needle swings on the ohmmeter scale.

17. Good.

18. Leaky.

19. Shorted.

20. Open.

 **Chapter Test**

13

Capacitance

Name: _____

Date: _____

Class: _____

Define the following technical terms.

1. Capacitance: _____

2. Dielectric: _____

3. Electrostatic field: _____

4. Plate: _____

5. Draw a block-diagram of a capacitor and label the following:
   a. Dielectric.
   b. Electrostatic field.
   c. Plate.

Fill in the blanks.

6. Aluminum foil is used to form the _____ of an electrolytic capacitor.

7. Air is used in some capacitors as the _____.

8. A capacitor opposes a change in _____.

9. When a capacitor is charging with a resistor in series, how does the resistor voltage compare to the capacitor voltage?

   _____

   _____

10. When a capacitor is discharging with a resistor in series, how does the resistor voltage compare to the capacitor voltage?

_____

_____

11. List three factors affecting capacitance.

_____

_____

_____

12. List three general types of capacitors.

_____

_____

_____

With each problem, write the formula, substitution, and answer.

13. Calculate the resultant of two capacitors connected in parallel, with values of 0.45 µF and 0.50 µF.

Formula: _____

Substitution: _____

Answer: _____

14. Calculate the resultant of two capacitors connected in series with values of 400 μF and 600 μF.

Formula: _____

Substitution: _____

Answer: _____

15. Calculate the voltages of a capacitive voltage divider with two capacitors connected in series with values of 1200 μF and 1400 μF. The applied voltage is 10 volts.

Formula: _____

Substitution: _____

V_{C_1}: _____

V_{C_2}: _____

16. List the steps to test a capacitor with an ohmmeter. _____

_____

_____

_____

For the following questions, an ohmmeter is used to test a capacitor. Use the observations of the ohmmeter needle to determine if the capacitor is good, shorted, open, or leaky.

17. _____The needle swings up towards zero ohms, then drops back towards infinity.

18. _____The needle goes to the middle of the scale and stays constant.

19. _____The needle reads almost zero ohms.

20. _____The needle reads infinity.

Time Constants and Waveshaping

OBJECTIVES

After studying this chapter, students should be able to:
- Define time constants as related to inductors and capacitors.
- Calculate the values of time constants.
- Plot a universal time constant curve and use it to predict the performance of time constant circuits.
- Calculate the performance of time constant circuits.
- Plot the output waveforms of waveshaping circuits with different time constants.
- Use waveshaping circuits in filter applications.
- Apply time constants to circuits requiring high voltage or high current.

INSTRUCTIONAL MATERIALS

Text: Pages 385–424
Test Your Knowledge Questions, Pages 422–423
Study Guide: Pages 111–126
Laboratory Manual: Pages 131–156

ANSWERS TO TEXTBOOK

Test Your Knowledge, Pages 422–423
1. 95%
2. Actual time is what passes by in life. A time constant is the time required to reach 63.2% of the remaining charge or discharge of an inductive or capacitive circuit.
3. 5 time constants to reach full charge.
4. 5 time constants to reach full discharge.
5. 63.2%
6. 86.5%
7. 95%
8. $\tau = \dfrac{L}{R}$

9. $\tau = R \times C$

10. $\tau = 0.5$ ms
Full charge = 2.5 ms

11. $\tau = 0.3$ s
Full charge = 1.5 s

12. One time constant = 0.25 μs
0.65 μs = 2.6 time constants
93% full charge (allow for reading graph)
Full charge current = 7.14 mA
Answer: $i = 6.64$ mA

13. One time constant = 0.2 s
0.3 s = 1.5 time constants
21% charge remaining (allow for reading graph)
Full charge current = 1 A
Answer: $i = 0.21$ A

14. One time constant = 0.5 s
1.2 s = 2.4 time constants
91% full charge (allow for reading graph)
Full charge voltage = 10 V
Answer: $v = 9.1$ V

15. One time constant = 2.5 s
2 s = 0.8 time constants
40% charge remaining (allow for reading graph)
Full charge voltage = 10 V
Answer: $v = 4$ V

16. One time constant = 0.2 ms
0.27 ms = 1.35 time constants
74.1% full charge (calculated with exponential formula)
Full charge current = 4 mA
Answer: $i = 2.96$ mA

17. One time constant = 5 ms
     7.1 ms = 1.42 time constants
     24.2% charge remaining (calculated with expo-
         nential formula)
     Full charge current = 0.2 A
     Answer: i = 48.4 mA

18. One time constant = 3 ms
     10.2 ms = 3.4 time constants
     96.7% full charge (calculated with exponential
         formula)
     Full charge voltage = 10 V
     Answer: v = 9.67 V

19. One time constant = 0.03 s
     0.12 s = 4 time constants
     1.83% charge remaining (calculated with expo-
         nential formula)
     Full charge voltage = 10 V
     Answer: v = 0.183 V

20. Refer to figures 14-11 and 14-13.
21. Refer to figure 14-11.
22. Refer to figures 14-16 and 14-18.
23. Refer to figure 14-16.
24. Leading edge.
25. Top and bottom (the dc voltage portion of the
     square wave).
26. Full charge current = 0.075 A
     Discharge voltage = 7500 V
27. I = 2.4 A

ANSWERS TO STUDY GUIDE

Pages 111–126
1. i.
2. l.
3. j.
4. f.
5. h.
6. k.
7. d.
8. a.
9. e.
10. m.
11. g.
12. b.
13. c.
14. b.
15. a.
16. a.
17. b.
18. d.
19. a.
20. a.
21. b.

22. a. One time constant:

     Formula: $\tau = \dfrac{L}{R}$

     Substitution: $\tau = \dfrac{600 \text{ mH}}{1000 \text{ }\Omega}$

     Answer: τ = 0.6 ms

    b. Time for full charge:
        Formula: Full charge = $5 \times \tau$
        Substitution: $5\tau = 5 \times 0.6$ ms
        Answer: 5τ = 3 ms

23. a. One time constant:
        Formula: $\tau = R \times C$
        Substitution: $\tau = 45 \text{ k}\Omega \times 20 \text{ }\mu\text{F}$
        Answer: τ = 0.9 s

    b. Time for full charge:
        Formula: Full charge = $5 \times \tau$
        Substitution: $5\tau = 5 \times 0.9$ s
        Answer: 5τ = 4.5 s

24. a. One time constant:

        Formula: $\tau = \dfrac{L}{R}$

        Substitution: $\tau = \dfrac{50 \text{ }\mu\text{H}}{200 \text{ }\Omega}$

        Answer: τ = 0.25 μs

b. Instantaneous time changed to time constants:

Formula: $T = \dfrac{t}{\tau}$

Substitution: $T = \dfrac{0.75\ \mu s}{0.25\ \mu s}$

Answer: $T = 3$ time constants

c. Percentage of full charge, from graph:
95.0% full charge (allow for reading graph)

d. Full charge current:

Formula: $I = \dfrac{V}{R}$

Substitution: $I = \dfrac{10\ V}{200\ \Omega}$

Answer: $I = 50$ mA

e. Instantaneous current:
Formula: $i = \% \times I$
Substitution: $i = 95\% \times 50$ mA
Answer: $i = 47.5$ mA

25. a. One time constant:

Formula: $\tau = \dfrac{L}{R}$

Substitution: $\tau = \dfrac{5\ H}{50\ \Omega}$

Answer: $\tau = 0.1$ s

b. Instantaneous time changed to time constants:

Formula: $T = \dfrac{t}{\tau}$

Substitution: $T = \dfrac{0.15\ s}{0.1\ s}$

Answer: $T = 1.5$ time constants

c. Percentage of full discharge, from graph:
23% charge remaining (allow for reading
graph)

d. Full charge current:

Formula: $I = \dfrac{V}{R}$

Substitution: $I = \dfrac{10\ V}{50\ \Omega}$

Answer: $I = 0.2$ A

e. Instantaneous current:
Formula: $i = \% \times I$
Substitution: $i = 23\% \times 0.2$ A
Answer: $i = 0.046$ A

26. a. One time constant:
Formula: $\tau = R \times C$
Substitution: $\tau = 1000\ \Omega \times 500\ \mu F$
Answer: $\tau = 0.5$ s

b. Instantaneous time changed to time constants:

Formula: $T = \dfrac{t}{\tau}$

Substitution: $T = \dfrac{1.2\ s}{0.5\ s}$

Answer: $T = 2.4$ time constants

c. Percentage of full charge, from graph:
90% full charge (allow for reading graph)

d. Full charge voltage:
Formula: No formula needed.
Substitution:
Full charge voltage = applied voltage
Answer: $V = 10$ V

e. Instantaneous capacitive voltage:
Formula: $v = \% \times V$
Substitution: $v = 90\% \times 10$ V
Answer: $v = 9$ V

27. a. One time constant:
Formula: $\tau = R \times C$
Substitution: $\tau = 10\ k\Omega \times 150\ \mu F$
Answer: $\tau = 1.5$ s

b. Instantaneous time changed to time constants:

Formula: $T = \dfrac{t}{\tau}$

Substitution: $T = \dfrac{2 \text{ s}}{1.5 \text{ s}}$

Answer: $T = 1.3$ time constants

c. Percentage of full discharge, from graph: 28% charge remaining (allow for reading graph)

d. Full charge voltage:
Formula: No formula needed.
Substitution:
    Full charge voltage = applied voltage
Answer: $V = 10$ V

e. Instantaneous capacitive voltage:
Formula: $v = \% \times V$
Substitution: $v = 28\% \times 10$ V
Answer: $v = 2.8$ V

28. a. One time constant:

Formula: $\tau = \dfrac{L}{R}$

Substitution: $\tau = \dfrac{0.5 \text{ H}}{2.5 \text{ k}\Omega}$

Answer: $\tau = 0.2$ ms

b. Instantaneous time changed to time constants:

Formula: $T = \dfrac{t}{\tau}$

Substitution: $T = \dfrac{0.27 \text{ ms}}{0.2 \text{ ms}}$

Answer: $T = 1.35$ time constants

c. Percentage of full charge:
Formula: $\% = (1 - e^{-T}) \times 100\%$
Substitution: $\% = (1 - e^{-1.35}) \times 100\%$
Answer: $\% = 74.1\%$

d. Full charge current:

Formula: $I = \dfrac{V}{R}$

Substitution: $I = \dfrac{10 \text{ V}}{2.5 \text{ k}\Omega}$

Answer: $I = 4$ mA

e. Instantaneous current:
Formula: $i = \% \times I$
Substitution: $i = 74.1\% \times 4$ mA
Answer: $i = 2.96$ mA

29. a. One time constant:

Formula: $\tau = \dfrac{L}{R}$

Substitution: $\tau = \dfrac{340 \text{ mH}}{170 \text{ }\Omega}$

Answer: $\tau = 2$ ms

b. Instantaneous time changed to time constants:

Formula: $T = \dfrac{t}{\tau}$

Substitution: $T = \dfrac{6.3 \text{ ms}}{2 \text{ ms}}$

Answer: $T = 3.15$ time constants

c. Percentage of full discharge:
Formula: $\% = (e^{-T}) \times 100\%$
Substitution: $\% = (e^{-3.15}) \times 100\%$
Answer: $\% = 4.3\%$ charge remaining

d. Full charge current:

Formula: $I = \dfrac{V}{R}$

Substitution: $I = \dfrac{10 \text{ V}}{170 \text{ }\Omega}$

Answer: $I = 58.8$ mA

e. Instantaneous current:
Formula: $i = \% \times I$
Substitution: $i = 4.3\% \times 58.8$ mA
Answer: $i = 2.53$ mA

30. a. One time constant:
Formula: $\tau = R \times C$
Substitution: $\tau = 500\ \Omega \times 25\ \mu F$
Answer: $\tau = 12.5$ ms

b. Instantaneous time changed to time constants:

Formula: $T = \dfrac{t}{\tau}$

Substitution: $T = \dfrac{10\ ms}{12.5\ ms}$

Answer: $T = 0.8$ time constants

c. Percentage of full charge:
Formula: $\% = (1 - e^{-T}) \times 100\%$
Substitution: $\% = (1 - e^{-0.8}) \times 100\%$
Answer: $\% = 55.1\%$

d. Full charge voltage:
Formula: No formula needed.
Substitution:
    Full charge voltage = applied voltage
Answer: $V = 10$ V

e. Instantaneous capacitive voltage:
Formula: $v = \% \times V$
Substitution: $v = 55.1\% \times 10$ V
Answer: $v = 5.51$ V

31. a. One time constant:
Formula: $\tau = R \times C$
Substitution: $\tau = 10\ k\Omega \times 250\ \mu F$
Answer: $\tau = 2.5$ s

b. Instantaneous time changed to time constants:

Formula: $T = \dfrac{t}{\tau}$

Substitution: $T = \dfrac{3\ s}{2.5\ s}$

Answer: $T = 1.2$ time constants

c. Percentage of full discharge:
Formula: $\% = (e^{-T}) \times 100\%$
Substitution: $\% = (e^{-1.2}) \times 100\%$
Answer: $\% = 30.1\%$ charge remaining

d. Full charge voltage:
Formula: No formula needed.
Substitution:
    Full charge voltage = applied voltage
Answer: $V = 10$ V

e. Instantaneous capacitive voltage:
Formula: $v = \% \times V$
Substitution: $v = 30.1\% \times 10$ V
Answer: $v = 3.01$ V

32. Refer to textbook figures 14-11 and 14-13.
33. Refer to textbook figure 14-11.
34. Refer to textbook figures 14-16 and 14-18.
35. Refer to textbook figure 14-16.
36. a.
37. b.

38. Full charge current, use Ohm's law with charge resistor and applied voltage.
    $I = 0.15$ A

Discharge voltage, use Ohm's law with charge current and discharge resistance.
    $V = 3750$ V

39. Use Ohm's law with charge voltage and discharge resistance.
    $I = 0.6$ A

ANSWERS TO CHAPTER TEST IN THE INSTRUCTOR'S MANUAL

Pages 181–193
1. Filter: A circuit that offers varying degrees of opposition to different frequencies.
2. One time constant: The length of time needed to reach 63.2% of full charge or discharge.
3. Inductive kick: The high voltage produced when the discharge path for an inductor has extremely high resistance.
4. Integrator: An application of a time constant circuit. If the input square wave is positive, the output has a shape the same as the universal time constant curve. The steepness depends on the time constant.
5. Long time constant circuit: A circuit with its full charge/discharge time greater than the time of one pulse width.

6.

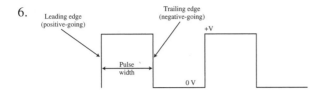

7. b.
8. a.
9. d.
10. c.
11. a.
12. b.
13. e.
14. f.

15. a. One time constant:

Formula: $\tau = \dfrac{L}{R}$

Substitution: $\tau = \dfrac{200 \text{ mH}}{4000 \ \Omega}$

Answer: $\tau = 50 \ \mu s$

b. Full charge:
Formula: Full = $5 \times \tau$
Substitution: Full = $5 \times 50 \ \mu s$
Answer: Full = $250 \ \mu s$

16. a. One time constant:
Formula: $\tau = R \times C$
Substitution: $\tau = 80 \text{ k}\Omega \times 40 \ \mu F$
Answer: $\tau = 3.2$ s

b. Full charge:
Formula: Full = $5 \times \tau$
Substitution: Full = 5×3.2 s
Answer: Full = 16 s

17. a. Current when fully charged:

Formula: $I = \dfrac{V}{R}$

Substitution: $I = \dfrac{10 \text{ V}}{500 \ \Omega}$

Answer: $I = 0.02$ A

b. Instantaneous current at 1.5 time constants:
Formula: i at 1.5 time constants = $77\% \times I$ full
Substitution: $i = 0.77 \times 0.02$ A
Answer: $i = 0.0154$ A

18. a. Capacitive voltage when fully charged:
10 V (No calculation needed. Full charge
voltage equals the applied voltage.)

b. Instantaneous capacitive voltage after 2.5
time constants:
Formula: $v = \% \times V_A$
Substitution: $v = 0.08 \times 10$ V
Answer: $v = 0.8$ V

19. a. Percentage of full charge:
Formula: $\% = (1 - e^{-T}) \times 100\%$
Substitution: $\% = (1 - e^{-0.85}) \times 100\%$
Answer: $\% = 57.3\%$

b. Current when fully charged:

Formula: $I = \dfrac{V}{R}$

Substitution: $I = \dfrac{10 \text{ V}}{1.6 \text{ k}\Omega}$

Answer: $I = 6.25$ mA

c. Formula: $i = \% \times I_{max}$
Substitution: $i = 0.573 \times 6.25$ mA
Answer: $i = 3.58$ mA

20. a. One time constant:
Formula: $\tau = R \times C$
Substitution: $\tau = 5 \text{ k}\Omega \times 450 \ \mu F$
Answer: $\tau = 2.25$ s

b. Instantaneous time changed to time constants:

Formula: $T = \dfrac{t}{\tau}$

Substitution: $T = \dfrac{2.0 \text{ s}}{2.25 \text{ s}}$

Answer: $T = 0.889$

c. Percentage of full discharge:
Formula: $\% = (e^{-T}) \times 100\%$
Substitution: $\% = (e^{-0.889}) \times 100\%$
Answer: $\% = 41.1\%$

d. Capacitive voltage when fully charged:
10 V (No calculation needed. Full charge
voltage equals the applied voltage.)

e. Instantaneous capacitive voltage:
Formula: $v = \% \times V_A$
Substitution: $v = 0.411 \times 10$ V
Answer: $v = 4.11$ V

21. This is a medium time constant circuit.
5 time constants = pulse width.

a. Draw the circuit diagram of an RC integrator.

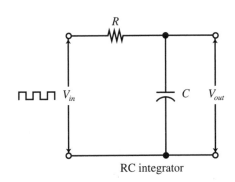

RC integrator

b. Draw the square wave input.

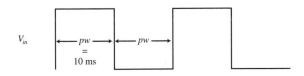

c. Draw the output waveform.

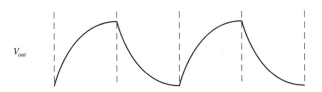

22. This is a medium time constant circuit.
5 time constants = pulse width.

a. Draw the circuit diagram of an RC differentiator.

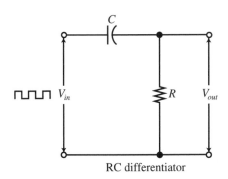

RC differentiator

b. Draw the square wave input.

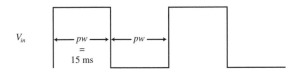

c. Draw the output waveform.

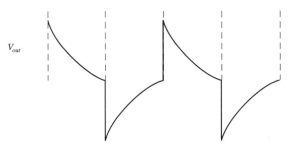

23. b.

24. Charge current = V/R = 20V/50 Ω = 0.4 A
Discharge voltage:
Formula: $V = I \times R$
Substitution: $V = 0.4$ A $\times$ 50 kΩ
Answer: $V = 20$ kV

25. Formula: $I = \dfrac{V}{R}$

Substitution: $I = \dfrac{9 \text{ V}}{3 \text{ }\Omega}$

Answer: $I = 3$ A

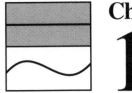

Chapter Test

14

Name: _____

Date: _____

Class: _____

Time Constants and Waveshaping

Define the following technical terms.

1. Filter: _____

2. One time constant: _____

3. Inductive kick: _____

4. Integrator: _____

5. Long time constant circuit: _____

6. Draw a square wave and label the following:
   a. Trailing edge.
   b. Leading edge.
   c. Pulse width.

Select the best answer.

_____ 7. One time constant is:
   a. the amount of time for every circuit.
   b. the time to reach a voltage equal to 63.2% of full charge.
   c. the time to reach a voltage equal to $1/5$ of full charge.
   d. the time to reach full charge or discharge.

_____ 8. How many time constants does it take to reach full discharge?
   a. 5
   b. 63.2
   c. 1
   d. 10

_____   9. The percentage of full charge or discharge reached in two time constants is _____.
            a. 20%
            b. 40%
            c. 63.2%
            d. 86.5%

_____   10. The percentage of full charge or discharge is reached in three time constants is _____.
            a. 63.2%
            b. 86.5%
            c. 95.0%
            d. 99.7%

For questions 11 to 14, select a formula from the following list that is used to solve each given application.
            a. $\tau = L/R$
            b. $\tau = R \times C$
            c. $\tau = L \times R$
            d. $\tau = R/C$
            e. $\% = (1 - e^{-T}) \times 100\%$
            f. $\% = (e^{-T}) \times 100\%$

11. Find one time constant in an inductive circuit. _____

12. Find one time constant in a capacitive circuit. _____

13. Find the instantaneous percentage of full charge. _____

14. Find the instantaneous percentage of full discharge. _____

With each problem, write the formula, substitution, and answer.

15. Calculate the time for one time constant and the time to reach full charge in a circuit with a 4000 ohm resistor in series with a 200 mH inductor.

   a. One time constant:

   Formula: _____

   Substitution: _____

   Answer: _____

   b. Full charge:

   Formula: _____

   Substitution: _____

   Answer: _____

16. Calculate the time for one time constant and the time to reach full charge in a circuit with a 80 kilohm resistor in series with a 40 μF capacitor.

   a. One time constant:

   Formula: _____

   Substitution: _____

   Answer: _____

   b. Full charge:

   Formula: _____

   Substitution: _____

   Answer: _____

17. After 1.5 time constants, a circuit is shown to have 77% of full charge. The circuit is a 100 μH inductor in series with a 500 ohm resistor with 10 volts applied. Solve for the following:

a. Current when fully charged:

Formula: _____

Substitution: _____

Answer: _____

b. Instantaneous current at 1.5 time constants:

Formula: _____

Substitution: _____

Answer: _____

18. After 2.5 time constants, a circuit is shown to have discharged to 8% of full charge. The circuit is a 250 µF capacitor in series with a 20 kilohm resistor with 10 volts applied. Solve for the following:

a. Capacitive voltage when fully charged:

Formula: _____

Substitution: _____

Answer: _____

b. Instantaneous capacitive voltage after 2.5 time constants:

Formula: _____

Substitution: _____

Answer: _____

19. Use the exponential formula to find the instantaneous charge current after 0.85 time constants in a circuit with a 0.4 H inductor in series with a 1.6 kilohm resistor and 10 volts applied.

   a. Percentage of full charge:

      Formula: _____

      Substitution: _____

      Answer: _____

   b. Current when fully charged:

      Formula: _____

      Substitution: _____

      Answer: _____

c. Instantaneous current at 0.85 time constants:

Formula: _____

Substitution: _____

Answer: _____

20. Use the exponential formula to find the instantaneous discharge voltage after 2.0 seconds in a circuit with a 450 μF capacitor in series with a 5 kilohm resistor and 10 volts applied.

a. One time constant:

Formula: _____

Substitution: _____

Answer: _____

Name: _____

b. Instantaneous time changed to time constants:

Formula: _____

Substitution: _____

Answer: _____

c. Percentage of full discharge:

Formula: _____

Substitution: _____

Answer: _____

d. Capacitive voltage when fully charged:

Formula: _____

Substitution: _____

Answer: _____

e. Instantaneous capacitive voltage:

Formula: _____

Substitution: _____

Answer: _____

21. Draw the circuit diagram and plot the output waveform of an RC integrator waveshaping circuit with a 2 ms time constant and 10 ms pulse width.

    a. Draw the circuit diagram of an RC integrator.

    b. Draw the square wave input.

    c. Draw the output waveform.

22. Plot the output waveform of an RC differentiator waveshaping circuit with a 3 ms time constant and 15 ms pulse width.

   a. Draw the circuit diagram of an RC differentiator.

   b. Draw the square wave input.

   c. Draw the output waveform.

_____ 23. When a differentiator with a long time constant is used with a square wave, what part of the output wave-form is affected the most?
   a. Leading edge.
   b. Trailing edge.

24. An RL time constant circuit has an inductor with 50 ohms internal resistance and has 20 volts applied during the charge time. Calculate the voltage developed across a 50 kilohm discharge resistor.

Formula: _____

Substitution: _____

Answer: _____

25. An RC time constant circuit charges the capacitor to 9 volts. What is the current during discharge with a load resistor of 3 ohms?

Formula: _____

Substitution: _____

Answer: _____

Chapter 15

AC Waveforms

OBJECTIVES

After studying this chapter, students should be able to:
- Distinguish between dc and ac voltages.
- Describe the characteristics of an ac waveform.
- Analyze square waves: positive, negative, and symmetrical.
- Calculate the four basic methods of measuring voltage.
- Calculate the period and frequency of ac waveforms.
- Explain how a sine wave is produced from a generator.
- Calculate the instantaneous voltage values along a sine wave.
- Recognize phase shift, harmonic frequencies, and other characteristics of ac waveforms.
- Analyze basic ac circuits with resistive loads.
- Use an oscilloscope to observe and measure an ac waveform.

INSTRUCTIONAL MATERIALS

Text: Pages 425–460
Test Your Knowledge Questions, Pages 458–460
Study Guide: Pages 127–135
Laboratory Manual: Pages 157–165

ANSWERS TO TEXTBOOK

Test Your Knowledge, Pages 458–460

1. A waveform is the shape of the voltage over a period of time
2. Voltage: The ac waveform changes polarity, from a maximum positive to a maximum negative.
   Time: The ac waveform repeats itself periodically over time.
3. Amplitude is measured from maximum positive to maximum negative.
4. One cycle is measured from any point along a waveform to the point it repeats itself.
5. Period is measured in units of time.
6. Peak-to-peak is twice peak.
7. 40 volts.
8. 50 volts.
9. $t = 1/f$ (t stands for period and f for frequency)
10. $t = 0.667$ ms
11. $f = 1/t$
12. $f = 250$ Hz
13. $v = V_{max} \times \sin \theta$
14. $v = 86.6$ V
15. $\theta = 30°$
16. 270°
17. $\pi/4$ rad
18. rms
19. p-to-p = 80 V
    rms = 28.28 V
    avg = 25.44 V
20. peak = 120 V
    rms = 84.84 V
    avg = 76.32 V
21. peak = 42.4 V
    p-to-p = 84.8 V
    avg = 26.97 V
22. peak = 78.62 V
    p-to-p =157.24 V
    rms = 55.58 V
23. p-to-p = 22 V
    peak = 11 V
    rms = 7.777 V
    avg = 6.996 V
    period = 120 ms
    freq = 8.33 Hz

24. p-to-p = 34 V
     peak = 17 V
     rms = 12 V
     avg = 10.8 V
     period = 20 ms
     freq = 50 Hz
24. p-to-p = 50 V
     peak = 25 V
     rms = 17.68 V
     avg = 15.9 V
     period = 25 ms
     freq = 40 Hz

ANSWERS TO STUDY GUIDE

Pages 127–135

1. e.
2. a.
3. s.
4. b.
5. l.
6. m.
7. r.
8. d.
9. n.
10. f.
11. h.
12. i.
13. p.
14. q.
15. c.
16. o.
17. k.
18. j.
19. g.
20. a. Voltage.
    b. Time.
21. Maximum negative to maximum positive.
22. One point to where it repeats.
23. a.
24. c.
25. Formula: p-to-p = 2 × peak
    Substitution: p-to-p = 2 × 20 V
    Answer: p-to-p = 40 V
26. Formula: peak = p-to-p/2
    Substitution: peak = 100 V/2
    Answer: peak = 50 V
27. Formula: $t = 1/f$
    Substitution: $t = 1/1500$ Hz
    Answer: $t = 0.667$ ms

28. Formula: $f = 1/t$
    Substitution: $f = 1/4$ ms
    Answer: $f = 250$ Hz
29. Formula: $v = V_{max} \times \sin \theta$
    Substitution: $v = 100$ V $\times \sin 60°$
    Answer: $v = 86.6$ V
30. Formula: $v = V_{max} \times \sin \theta$
    Substitution: $\sin \theta = 50$ V/100 V
    Answer: $\theta = 30°$

31. Formula: degrees = $\dfrac{180°}{\pi \text{ rad}} \times$ rad

    Substitution: degrees = $\dfrac{180°}{\pi \text{ rad}} \times \dfrac{3\pi}{2}$

    Answer: degrees = 270°

32. Formula: rad = $\dfrac{\pi \text{ rad}}{180°} \times$ degrees

    Substitution: rad = $\dfrac{\pi \text{ rad}}{180°} \times 45°$

    Answer: rad = $\pi/4$ rad

33. c.
34. a. Formula: p-to-p = 2 × peak
        Substitution: p-to-p = 2 × 40 V
        Answer: p-to-p = 80 V
    b. Formula: rms = peak × 0.707
        Substitution: rms = 40 V × 0.707
        Answer: rms = 28.28 V
    c. Formula: avg = peak × 0.636
        Substitution: = 40 V × 0.636
        Answer: avg = 25.44 V
35. a. peak = 120 V
    b. rms = 84.84 V
    c. avg = 76.32 V
36. a. peak = 42.4 V
    b. p-to-p = 84.8 V
    c. avg = 26.97 V
37. a. peak = 78.6 V
    b. p-to-p = 157.2 V
    c. rms = 55.57 V
38. a. Number of divisions: 5
        Volts per division: 200 mV
        p-to-p = 1 V
    b. Formula: peak = p-to-p/2
        Substitution: peak = 1 V/2
        Answer: peak = 500 mV

c. Formula: rms = peak × 0.707
Substitution: rms = 500 mV × 0.707
Answer: rms = 353.5 mV

d. Formula: avg = peak × 0.636
Substitution: avg = 500 mV × 0.636
Answer: avg = 318 mV

e. Number of divisions: 3.6
Time per division: 100 ms
period = 360 ms

f. Formula: $f = 1/t$
Substitution: $f = 1/360$ ms
Answer: $f = 2.78$ Hz

39. a. Number of divisions: 4
Volts per division: 100 mV
p-to-p = 400 mV

b. Formula: peak = p-to-p/2
Substitution: peak = 400 mV/2
Answer: peak = 200 mV

c. Formula: rms = peak × 0.707
Substitution: rms = 200 mV × 0.707
Answer: rms = 141.4 mV

d. Formula: avg = peak × 0.636
Substitution: avg = 200 mV × 0.636
Answer: avg = 127.2 mV

e. Number of divisions: 6
Time per division: 1 ms
period = 6 ms

f. Formula: $f = 1/t$
Substitution: $f = 1/6$ ms
Answer: $f = 166.7$ Hz

40. a. Number of divisions: 2.8
Volts per division: 500 mV
p-to-p = 1.4 V

b. Formula: peak = p-to-p/2
Substitution: peak = 1.4 V/2
Answer: peak = 700 mV

c. Formula: rms = peak × 0.707
Substitution: rms = 700 mV × 0.707
Answer: rms = 494.9 mV

d. Formula: avg = peak × 0.636
Substitution: avg = 700 mV × 0.636
Answer: avg = 445.2 mV

e. Number of divisions: 4.8
Time per division: 100 μs
period = 480 μs

f. Formula: $f = 1/t$
Substitution: $f = 1/480$ μs
Answer: $f = 2083$ Hz

41. a. Number of divisions: 4
Volts per division: 50 mV
p-to-p = 200 mV

b. Formula: peak = p-to-p/2
Substitution: peak = 200 mV/2
Answer: peak = 100 mV

c. Formula: rms = peak × 0.707
Substitution: rms = 100 mV × 0.707
Answer: rms = 70.7 mV

d. Formula: avg = peak × 0.636
Substitution: avg = 100 mV × 0.636
Answer: avg = 63.6 mV

e. Number of divisions: 4
Time per division: 100 μs
period = 400 μs

f. Formula: $f = 1/t$
Substitution: $f = 1/400$ μs
Answer: $f = 2500$ Hz

ANSWERS TO CHAPTER TEST IN THE INSTRUCTOR'S MANUAL

Pages 199–201

1. Amplitude: Strength of an ac waveform. Amplitude is the height measured on an oscilloscope.

2. Cycle: On an ac waveform, it is measured from one point to the point where the wave repeats itself.

3. Period: The length of time for a waveform to complete one cycle.

4. Frequency: The number of cycles in one second.

5. Hertz: The unit measure of frequency.

6. amplitude, period (in any order)

7. equal

8. Cycle

9. time (seconds, milliseconds, microseconds, etc.)

10. sawtooth

For questions 11 through 22, the underlined values shown in this chart are the given values. The numbered answers are the solutions.

Peak	Peak To Peak	Average	RMS
40 V	(11.) 80 V	(12.) 25.44 V	(13.) 28.28 V
(14.) 45 V	90 V	(15.) 28.62 V	(16.) 31.82 V
(17.) 28.3 V	(18.) 56.6 V	18 V	(19.) 20.0
(20.) 50.92 V	(21.) 101.8 V	(22.) 32.38 V	36 V

23.   Formula: $t = \dfrac{1}{f}$

Substitution: $t = \dfrac{1}{2500 \text{ Hz}}$

Answer: $t = 0.4$ ms

24.   Formula: $f = \dfrac{1}{t}$

Substitution: $f = \dfrac{1}{8 \text{ ms}}$

Answer: $f = 125$ Hz

25.   Formula: $v = V_{max} \times \sin \theta$
Substitution: $v = 10 \text{ V} \times \sin 70$
Answer: $v = 9.4$ V

Chapter Test

15

AC Waveforms

Name: _____

Date: _____

Class: _____

Define the following technical terms.

1. Amplitude: _____

2. Cycle: _____

3. Period: _____

4. Frequency: _____

5. Hertz (Hz): _____

Fill in the blanks.

6. Two measurements commonly taken on ac waveforms are _____ and

   _____.

7. A symmetrical waveform has _____ values of positive and negative voltages.

8. _____ is the reciprocal of period.

9. The unit of measure for period is _____.

10. The ramp waveform is also called the _____ wave.

Complete the following chart. Convert the given values to those requested. All values are for a symmetrical sine wave.

Peak	Peak to Peak	Average	RMS
40 V	(11.)_____	(12.)_____	(13.)_____
(14.)_____	90 V	(15.)_____	(16.)_____
(17.)_____	(18.)_____	18 V	(19.)_____
(20.)_____	21.)_____	(22.)_____	36 V

23. Calculate the period of a 2500 Hz ac signal.

Formula: _____

Substitution: _____

Answer: _____

24. Calculate the frequency of a signal having a period of 8 ms.

Formula: _____

Substitution: _____

Answer: _____

25. Calculate the instantaneous value of a sine wave at 70° with a peak value of 10 volts.

Formula: _____

Substitution: _____

Answer: _____

Chapter 16
Transformers

OBJECTIVES

After studying this chapter, students should be able to:
- Define technical terms related to transformers.
- Describe how a transformer is constructed.
- Describe how magnetism is used to pass voltage from one coil to another.
- Perform calculations using the turns ratio, voltage ratio, and current ratio.
- Perform calculations of transformer power and current.
- Describe transformer losses and ways to reduce these losses.
- Calculate transformer efficiency.
- Calculate the reflected resistance of a transformer and perform impedance matching.
- Recognize the effects of transformer loading.
- Use a multimeter to troubleshoot a defective transformer.

INSTRUCTIONAL MATERIALS

Text: Pages 461–498
Test Your Knowledge Questions, Pages 497–498
Study Guide: Pages 137–142
Laboratory Manual: Pages 167–178

ANSWERS TO TEXTBOOK

Test Your Knowledge, Pages 497–498
1. mutual
2. The transformer has primary and secondary coils, usually with a core.
3. Current flowing in the primary develops a magnetic field which crosses the secondary coil. This magnetic field induces a voltage in the secondary.
4. The core holds the coils of wire and maximizes the transfer of magnetic energy.
5. Different output voltages can be obtained using the same coil.
6. Refer to figures 16-8 and 16-9.
7. By adjusting where the output tap makes contact with the coil.
8. To remove possible shock hazard by having a ground connection on the secondary.
9. 5 V, step-down
10. 200 V, step-up
11. 250 V, step-down
12. 1 V, step-up
13. 1:12, step-up
14. 10:1, step-down
15. 10 A, step-down
16. 12 A, step-down
17. 80%
18. 5333 W
19. 1.25 A
20. $R_P = 180\ \Omega$
21. $R_P = 720\ \Omega$
22. Copper, hysteresis, and eddy currents. Losses can be minimized by using lower frequencies and cores made of laminations.
23. DC voltages saturate the transformer's magnetic field.
24. Secondary voltage drops as load current is increased.
25. a. Shorted primary.
    b. Open secondary.
    c. Shorted secondary.

ANSWERS TO STUDY GUIDE

Pages 137–142
1. m.
2. h.
3. j.
4. b.
5. a.
6. g.

7. l.
8. k.
9. c.
10. f.
11. i.
12. d.
13. n.
14. e.
15. The transformer has primary and secondary coils, usually with a core.
16. Current flowing in the primary windings develops a magnetic field that crosses the secondary coil. The magnetic field induces a voltage in the secondary windings.
17. Hold the coils of wire in a firm position, maximizes the transfer of magnetic energy.
18. Different voltages can be obtained from the same coil.
19. a. Refer to textbook figure 16-8.
    b. Refer to textbook figure 16-9.
20. a.
21. d.

22. Formula: $\dfrac{N_P}{N_S} = \dfrac{V_P}{V_S}$

    Substitution: $\dfrac{20}{1} = \dfrac{1200 \text{ V}}{V_S}$

    Answer: $V_S = 60$ V, step-down

23. Formula: $\dfrac{N_P}{N_S} = \dfrac{V_P}{V_S}$

    Substitution: $\dfrac{1}{50} = \dfrac{10 \text{ V}}{V_S}$

    Answer: $V_S = 500$ V, step-up

24. Formula: $\dfrac{N_P}{N_S} = \dfrac{V_P}{V_S}$

    Substitution: $\dfrac{35}{1} = \dfrac{V_P}{5 \text{ V}}$

    Answer: $V_P = 175$ V, step-down

25. Formula: $\dfrac{N_P}{N_S} = \dfrac{V_P}{V_S}$

    Substitution: $\dfrac{1}{5} = \dfrac{V_P}{30 \text{ V}}$

    Answer: $V_P = 6$ V, step-up

26. Formula: $\dfrac{N_P}{N_S} = \dfrac{V_P}{V_S}$

    Substitution: $\dfrac{N_P}{N_S} = \dfrac{40 \text{ V}}{1000 \text{ V}}$

    Answer: turns ratio = 1:25, step-up

27. Formula: $\dfrac{N_P}{N_S} = \dfrac{V_P}{V_S}$

    Substitution: $\dfrac{N_P}{N_S} = \dfrac{120 \text{ V}}{24 \text{ V}}$

    Answer: turns ratio = 5:1, step-down

28. Formula: $P_P = P_S$
    Substitution: $120 \text{ V} \times I_P = 30 \text{ V} \times 4 \text{ A}$
    Answer: $I_P = 1$ A, step-down

29. Formula: $P_P = P_S$
    Substitution: $120 \text{ V} \times 0.5 \text{ A} = 30 \text{ V} \times I_S$
    Answer: $I_S = 2$ A, step-down

30. Formula: eff $= \dfrac{P_{out}}{P_{in}} \times 100\%$

    Substitution: eff $= \dfrac{600 \text{ W}}{800 \text{ W}} \times 100\%$

    Answer: eff = 75%

31. Formula: eff $= \dfrac{P_{out}}{P_{in}} \times 100\%$

    Substitution: $85\% = \dfrac{3000 \text{ W}}{P_{in}} \times 100\%$

    Answer: $P_{in} = 3529$ W

32. Formula: eff $= \dfrac{V_S \times I_S}{V_P \times I_P} \times 100\%$

    Substitution: $90\% = \dfrac{240 \text{ V} \times 5 \text{ A}}{120 \text{ V} \times I_P} \times 100\%$

    Answer: $I_P = 11.1$ A

33. Secondary voltage = 50 V
    Turns ratio = 2.4:1

    Formula: $\dfrac{R_P}{R_S} = \left(\dfrac{N_P}{N_S}\right)^2$

    Substitution: $\dfrac{R_P}{10\ \Omega} = \left(\dfrac{2.4}{1}\right)^2$

    Answer: $R_P = 57.6\ \Omega$

34. Formula: $\dfrac{R_P}{R_S} = \left(\dfrac{N_P}{N_S}\right)^2$

    Substitution: $\dfrac{R_P}{15\ \Omega} = \left(\dfrac{3}{1}\right)^2$

    Answer: $R_P = 135\ \Omega$

35. a. Copper.
    b. Hysteresis.
    c. Eddy currents.
36. DC voltages create a magnetic field that must be overcome by the ac in order to produce a voltage in the secondary. If the dc voltage is strong enough, it can saturate the transformer's magnetic field.
37. Secondary voltage drops as load current is increased.
38. b.
39. c.
40. d.

ANSWERS TO CHAPTER TEST IN THE INSTRUCTOR'S MANUAL

Pages 207–208

1. Autotransformer: A transformer that has only one coil used for both the primary and the secondary.
2. I^2R losses: Losses due to the copper wire in a transformer.
3. Eddy currents: Electrical current flowing within the core of an electromagnet.
4. Step-down transformer: A transformer with a smaller number of turns in secondary than the primary. It has a lower voltage in secondary than in the primary.
5. Step-up transformer: A transformer with a larger number of turns in secondary than the primary. It has a larger voltage in secondary than in the primary.

For questions 6 through 20, the underlined values shown in this chart are the given values. The numbered answers are the solutions.

Turns Ratio Pri: Sec	V_P	V_S	I_P	I_S	Step-up or Step-down
4:1	36 V	(6.) 9 V	1 A	(7.) 4 A	(8.) Step-down
1:6	10 V	(9.) 60 V	(10.) 3.0 A	0.5 A	(11.) Step-up
1:5	(12.) 20 V	100 V	0.2 A	(13.) 0.04 A	(14.) Step-up
10:1	(15.) 120 V	12 V	6 A	(16.) 60 A	(17.) Step-down
20:1	(18.) 480 V	24 V	(19.) 1 mA	20 mA	(20.) Step-down

For questions 21 through 23, the underlined values shown in this chart are the given values. The numbered answers are the solutions.

% Eff	Primary Power	Secondary Power
85%	500 W	(21.) 425 W
80%	(22.) 750 W	600 W
(23.) 75%	400 W	300 W

24. Copper losses.
    Hysteresis loss.
    Eddy currents.

25. Secondary voltage decreases as the load current increases from zero to maximum.

Transformers

Define the following technical terms.

1. Autotransformer: _____

2. I^2R losses: _____

3. Eddy currents: _____

4. Step-down transformer: _____

5. Step-up transformer: _____

Complete the following chart.

Turns Ratio Pri:Sec	V_P	V_S	I_P	I_S	Step-up or Step-down
4:1	36 V	(6.) _____	1 A	(7.) _____	(8.) _____
1:6	10 V	(9.) _____	(10.) _____	0.5 A	(11.) _____
1:5	(12.) _____	100 V	0.2 A	(13.) _____	(14.) _____
10:1	(15.) _____	12 V	6 A	(16.) _____	(17.) _____
20:1	(18.) _____	24 V	(19.) _____	20 mA	(20.) _____

Complete the following chart.

% Efficiency	Primary Power	Secondary Power
85%	500 W	(21.) _____
80%	(22.) _____	600 W
(23.) _____	400 W	300 W

24. State three types of transformer losses. _____

_____

_____

_____

25. What happens to the secondary voltage of a transformer when the load current is increased from zero to the maximum

rated current? _____

_____

_____

Electric Generators and Motors

OBJECTIVES

After studying this chapter, students should be able to:
- Define technical terms related to generators and motors.
- Describe the basic generator action to produce electricity.
- State the functions of the armature and fields of a generator.
- Name the parts of a commutator and describe their function.
- List and explain the methods of exciting a generator field.
- Describe the basic action of an electric motor.
- Describe the operation of a dc motor.
- Describe the operation of an ac motor.
- State the ratings for motors.
- Perform basic horsepower calculations.
- Perform basic torque calculations.

INSTRUCTIONAL MATERIALS

Text: Pages 499–522
Test Your Knowledge Questions, Page 521
Study Guide: Pages 143–145
Laboratory Manual: Pages 179–182

ANSWERS TO TEXTBOOK

Test Your Knowledge, Page 521
1. Generators use mechanical energy to produce electricity. Motors use electrical energy to produce mechanical force.
2. A coil of wire rotating through a magnetic field produces the electricity.
3. The armature is the rotating wire that spins past the magnetic fields.
   The fields generate the magnetic fields.

4. Split rings: Rotate with the armature.
   Brushes: Make contact with the split rings to bring current to an outside load.
5. a. Permanent magnets.
   b. Independent dc voltage.
   c. Self-excited shunt connection.
   d. Self-excited series connection.
   e. Self-excited compound connection.
   Student explanations will vary. See pages 504 through 507.
6. The magnetic field of the stator forces the rotor to spin.
7. DC voltages are used (or ac voltage with brushes and commutator) to energize the magnetic fields. The stator windings also receive current. The magnetic fields of the rotor and stator oppose and attract to create motor action.
8. AC voltages are used to energize a magnetically revolving stator field.
9. Horsepower, torque, voltage, frequency, output speed, duty cycle.
10. 2611 W
11. 2 hp
12. 6 ft-lb

ANSWERS TO STUDY GUIDE

Pages 211–213
1. a.
2. f.
3. o.
4. b.
5. p.
6. n.
7. l.
8. d.
9. i.
10. j.
11. q.

12. h.
13. r.
14. s.
15. k.
16. c.
17. m.
18. g.
19. t.
20. e.
21. c.
22. spins, coil of wire
23. is stationary, magnetic field
24. a. and b. Brushes, split rings (in any order).
25. a.   Permanent magnets.
    b.   Independent dc voltage.
    c.   Self-excited shunt connections.
    d.   Self-excited series connections.
    e.   Self-excited compound connections.
26. d.
27. c.
28. a.
29. a. and b. Horsepower, torque (in any order).
30. a. and b. Voltage, frequency (in any order).

31. Formula: mechanical power =

    $$\frac{746 \text{ watts}}{1 \text{ horsepower}} \times \text{motor rating (hp)}$$

    Substitution: power = $\frac{746 \text{ watts}}{1 \text{ horsepower}} \times 5$ hp

    Answer: power = 3730 W

32. Formula: power =

    $$\frac{1 \text{ horsepower}}{746 \text{ watts}} \times \text{motor rating (watts)}$$

    Substitution: power = $\frac{1 \text{ horsepower}}{746 \text{ watts}} \times 3730$ watts

    Answer: power = 5 hp

33. Formula: torque = force × distance
    Substitution: torque = 30 ounces × 40 inches
    Answer: torque = 1200 ounce-inches

ANSWERS TO CHAPTER TEST IN THE INSTRUCTOR'S MANUAL

Pages 211–213

1. armature
2. field
3. slip ring
4. brush
5. commutator
6. shunt generator
7. rotor
8. stator
9. induction motor
10. synchronous motor
11. centrifugal switch
12. shading ring
13. horsepower
14. torque
15. Duty cycle
16. c.
17. d.

18. Formula: power = $\dfrac{746 \text{ W}}{1 \text{ hp}} \times \text{motor rating (hp)}$

    Substitution: power = $\dfrac{746 \text{ W}}{1 \text{ hp}} \times 1.5$ hp

    Answer: power = 1119 W

19. Formula: power = $\dfrac{746 \text{ W}}{1 \text{ hp}} \times \text{motor rating (hp)}$

    Rearranging: motor rating = $\dfrac{1 \text{ hp}}{746 \text{ W}} \times \text{power}$

    Substitution: motor rating = $\dfrac{1 \text{ hp}}{746 \text{ W}} \times 2238$ W

    Answer: motor rating = 3 hp

20. Formula: torque = force × distance
    Substitution: torque = 50 oz × 100 in
    Answer: torque = 5000 oz-in

 **Chapter Test**

17

Name: _____

Date: _____

Class: _____

Electric Generators and Motors

Choose the best word or phrases from the following list to fill in the blanks in questions 1 through 15.

armature	rotor slip
brush	series generator
centrifugal switch	shading ring
compound generator	shunt generator
duty cycle	slip ring
field	split ring commutator
horsepower	stator
induction motor	synchronous motor
motor action	synchronous speed
rotor	torque

1. The _____ is the coil of wire in a generator spinning past the magnetic fields.

2. The _____ is the section of a generator providing magnetism.

3. In an ac generator, the _____ is the point where an armature wire is attached.

4. A(n) _____ is used in ac and dc generators and motors to connect an outside wire to the spinning armature.

5. The ac sine wave produced in a generator is changed to a pulsating dc voltage by the _____.

6. The _____ field winding is connected in parallel with the armature winding.

7. The _____ is the part of a motor that spins.

8. The _____ is the stationary magnetic field of a motor.

9. An ac motor with a magnetically revolving stator field is called a _____.

10. A(n) _____ matches its speed to the ac line frequency.

11. The device located on the rotor, called the _____, is used to disconnect the starting winding from the circuit.

12. The _____ is a copper ring around the notch cut into the stator field.

13. One _____ is equal to 746 watts of electrical power.

14. The rotational force of a motor is called _____.

15. _____ is the amount of time a motor is on compared to the time off.

Select the best answer.

_____ 16. A generator produces electricity by:
    a. a magnetic field spinning past another magnetic field.
    b. two coils of wire magnetically coupled.
    c. a wire spinning through a magnetic field.
    d. attraction and repulsion of magnetic fields.

_____ 17. An electric motor spins because of:
    a. an electric charge inside a magnetic field.
    b. two coils of wire magnetically coupled.
    c. a wire spinning through a magnetic field.
    d. attraction and repulsion of magnetic fields.

With each problem, write the formula, substitution, and answer.

18. Calculate the wattage of a 1.5 horsepower electric motor.

Formula: _____

Substitution: _____

Answer: _____

19. Calculate the horsepower of a motor if it has an electric power rating of 2238 watts.

Formula: _____

Substitution: _____

Answer: _____

20. Calculate the torque exerted by an electric motor lifting 50 ounces a distance of 100 inches.

Formula: _____

Substitution: _____

Answer: _____

Chapter 18

Inductive Reactance and Impedance

OBJECTIVES

After studying this chapter, students should be able to:
- Define reactance and impedance.
- Apply Ohm's law to inductive reactive circuits.
- Calculate inductive reactance.
- Examine the variables in the inductive reactance formula.
- Calculate the total reactance of series and parallel circuits.
- Examine the phase shift of current vs voltage in a purely inductive circuit.
- Construct phasor diagrams to show relationships in a series circuit with both reactance and resistance.
- Calculate impedance in a series circuit.
- Compare the effects of a larger and smaller X_L as compared to resistance in a series circuit.
- Construct phasor diagrams to show relationships in a parallel circuit with both reactance and resistance.
- Calculate impedance in a parallel circuit.
- Compare the effects of a larger and smaller X_L as compared to resistance in a parallel circuit.
- Measure phase angle with an oscilloscope.

INSTRUCTIONAL MATERIALS

Text: Pages 523–564
   Test Your Knowledge Questions, Pages 560–564
Study Guide: Pages 147–158
Laboratory Manual: 183–201

ANSWERS TO TEXTBOOK

Test Your Knowledge, Pages 560–564

1. Reactance is measured indirectly. It is found by measuring current and the applied voltage. Ohm's law is then used to calculate the value.

2. $X_L = 20\ \Omega$
3. $X_L = 628\ \Omega$
4. An increase in frequency will increase X_L.
5. A decrease in the value of inductance will decrease X_L.
6. $X_{L_T} = 150\ \Omega$
7. $X_{L_T} = 14.1\ \Omega$
8. Refer to figure 18-13.
9. a. $X_L = 80\ \Omega$
   b. $Z = 113\ \Omega$
   c. Student sketch of impedance triangle.
   d. $\theta = 45°$
   e. $\phi = 45°$
   f. $I = 0.442$ A
   g. $V_R = 35.4$ V
      $V_L = 35.4$ V
   h. Student sketch of voltage triangle.
10. a. $Z = 104\ \Omega$
   b. Student sketch of impedance triangle.
   c. $\theta = 73.3°$
   d. $\phi = 16.7°$
   e. $I = 0.577$A
   f. $V_R = 17.3$ V
      $V_L = 57.7$ V
   g. Student sketch of voltage triangle.
11. a. $Z = 104\ \Omega$
   b. Student sketch of impedance triangle.
   c. $\theta = 16.7°$
   d. $\phi = 73.3°$
   e. $I = 0.577$A
   f. $V_R = 57.7$ V
      $V_L = 17.3$ V
   g. Student sketch of voltage triangle.

12. a. $X_L = 60\ \Omega$
     b. $I_R = 1.33\ \text{A}$
     c. $I_L = 1.33\ \text{A}$
     d. $I_T = 1.88\ \text{A}$
     e. Student sketch of current triangle.
     f. $\theta = -45°$
     g. $Z = 42.5\ \Omega$

13. a. $I_R = 20\ \text{mA}$
     b. $I_L = 40\ \text{mA}$
     c. $I_T = 44.7\ \text{mA}$
     d. Student sketch of current triangle.
     e. $\theta = -63.4°$
     f. $Z = 224\ \Omega$

14. a. $I_R = 40\ \text{mA}$
     b. $I_L = 20\ \text{mA}$
     c. $I_T = 44.7\ \text{mA}$
     d. Student sketch of current triangle.
     e. $\theta = -26.6°$
     f. $Z = 224\ \Omega$

15. phase shift = 48°

ANSWERS TO STUDY GUIDE

Pages 147–158

1. i.
2. b.
3. a.
4. g.
5. f.
6. e.
7. d.
8. c.
9. h.
10. Formula: $X_L = V/I$
    Substitution: $X_L = 80\ \text{V}/4\ \text{A}$
    Answer: $X_L = 20\ \Omega$
11. Formula: $I = V/R$
    Substitution: $I = 25\ \text{V}/500\ \Omega$
    Answer: $I = 50\ \text{mA}$
12. Formula: $X_L = 2\pi f L$
    Substitution: $X_L = 2\pi \times 100\ \text{Hz} \times 200\ \text{mH}$
    Answer: $X_L = 126\ \Omega$
13. Formula: $X_L = 2\pi f L$
    Substitution: $150\ \Omega = 2\pi \times 50\ \text{Hz} \times L$
    Answer: $L = 0.477\ \text{H}$
14. Formula: $X_L = 2\pi f L$
    Substitution: $5\ \text{k}\Omega = 2\pi \times f \times 0.4\ \text{H}$
    Answer: $f = 1989\ \text{Hz}$

15. An increase in frequency will increase X_L.
16. A decrease in the value of inductance will decrease X_L.
17. Formula: $X_{L_T} = X_{L_1} + X_{L_2} + X_{L_3} + \ldots X_{L_N}$
    Substitution: $X_{L_T} = 50\ \Omega + 20\ \Omega + 60\ \Omega$
    Answer: $X_{L_T} = 130\ \Omega$

18. Formula: $\dfrac{1}{X_{L_T}} = \dfrac{1}{X_{L_1}} + \dfrac{1}{X_{L_2}} + \dfrac{1}{X_{L_3}} + \ldots \dfrac{1}{X_{L_N}}$
    Substitution: $\dfrac{1}{X_{L_T}} = \dfrac{1}{30\ \Omega} + \dfrac{1}{70\ \Omega} + \dfrac{1}{40\ \Omega}$
    Answer: $X_{L_T} = 13.8\ \Omega$

19. Refer to textbook figure 18-13.
20. a. Formula: $X_L = 2\pi f L$
        Substitution: $X_L = 2\pi \times 400\ \text{Hz} \times 0.8\ \text{H}$
        Answer: $X_L = 2011\ \Omega$

    b. Formula: $Z = \sqrt{X_L^2 + R^2}$
        Substitution: $Z = \sqrt{(2011\ \Omega)^2 + (2\ \text{k}\Omega)^2}$
        Answer: $Z = 2836\ \Omega$

    c. Student sketch of impedance triangle.

    d. Formula: $\theta = \tan^{-1} \dfrac{X_L}{R}$
        Substitution: $\theta = \tan^{-1} \dfrac{2011\ \Omega}{2\ \text{k}\Omega}$
        Answer: $\theta = 45°$

    e. Formula: $\phi = 90° - \theta$
        Substitution: $\phi = 90° - 45°$
        Answer: $\phi = 45°$
    f. Formula: $I = V/Z$
        Substitution: $I = 40\ \text{V}/2836\ \Omega$
        Answer: $I = 14.1\ \text{mA}$
    g. Formula: $V_R = I \times R$ and $V_L = I \times X_L$
        Substitution for V_R: $V_R = 14.1\ \text{mA} \times 2\ \text{k}\Omega$
        Answer: $V_R = 28.2\ \text{V}$
        Substitution for V_L: $V_L = 14.1\ \text{mA} \times 2011\ \Omega$
        Answer: $V_L = 28.4\ \text{V}$
    h. Student sketch of voltage triangle.

21. a. $Z = 510\ \Omega$
     b. Student sketch of impedance triangle.
     c. $\theta = 78.7°$
     d. $\phi = 11.3°$
     e. $I = 78.4$ mA
     f. $V_R = 7.84$ V
         $V_L = 39.2$ V
     g. Student sketch of voltage triangle.

22. a. $Z = 509\ \Omega$
     b. Student sketch of impedance triangle.
     c. $\theta = 11.3°$
     d. $\phi = 78.7°$
     e. $I = 78.4$ mA
     f. $V_R = 39.2$ V
         $V_L = 7.84$ V
     g. Student sketch of voltage triangle.

23. a. Formula: $X_L = 2\pi fL$
         Substitution: $X_L = 2\pi \times 5$ kHz $\times 38.2$ mH
         Answer: $X_L = 1.2$ kΩ
     b. Formula: $I_R = V/R$
         Substitution: $I_R = 60$ V/1.2 kΩ
         Answer: $I_R = 50$ mA
     c. Formula: $I_L = V/X_L$
         Substitution: $I_L = 60$ V/1.2 kΩ
         Answer: $I_L = 50$ mA

     d. Formula: $I_T = \sqrt{I_L{}^2 + I_R{}^2}$
         Substitution: $I_T = \sqrt{50\ \text{mA} + 50\ \text{mA}}$
         Answer: $I_T = 70.7$ mA

     e. Student sketch of current triangle.

     f. Formula: $\theta = \tan^{-1} \dfrac{-I_L}{I_R}$

         Substitution: $\theta = \tan^{-1} \dfrac{-50\ \text{mA}}{50\ \text{mA}}$

         Answer: $\theta = -45°$

     g. Formula: $Z = V_a/I_T$
         Substitution: $Z = 60$ V/70.7 mA
         Answer: $Z = 849\ \Omega$

24. a. $I_R = 6$ A
     b. $I_L = 1$ A
     c. $I_T = 6.08$ A
     d. Student sketch of current triangle.
     e. $\theta = -9.5°$
     f. $Z = 9.87\ \Omega$

25. a. $I_R = 1$ A
     b. $I_L = 6$ A
     c. $I_T = 6.08$ A
     d. Student sketch of current triangle.
     e. $\theta = -80.5°$
     f. $Z = 9.87\ \Omega$

26. a. $360° = 10$ divisions
     b. 1 division $= 36°$
     c. Difference between waveforms $= -1.2$ divisions
     d. $\theta = -43.2°$

27. a. $360° = 8$ divisions
     b. 1 division $= 45°$
     c. Difference between waveforms $= -0.6$ divisions
     d. $\theta = -27°$

28. a. $360° = 8$ divisions
     b. 1 division $= 45°$
     c. Difference between waveforms $= -1.2$ divisions
     d. $\theta = -43.2°$

ANSWERS TO CHAPTER TEST IN THE INSTRUCTOR'S MANUAL

Pages 221–230

1. Impedance: The total ac resistance of a circuit containing both resistance and reactance.
2. Inductive reactance: The ac resistance of an inductor.
3. Phase angle: The time difference between the applied voltage and the resistor voltage.
4. Phasor: A vector used to show the magnitude and direction of an electrical quantity.
5. Pythagorean theorem: A mathematical formula used to calculate the length of the hypotenuse of a right triangle: $a = \sqrt{b^2 + c^2}$.

6. Formula: $X_L = \dfrac{V}{I}$

    Substitution: $X_L = \dfrac{100\ \text{V}}{2\ \text{A}}$

    Answer: $X_L = 50\ \Omega$

7. Formula: $I = \dfrac{V}{X_L}$

Substitution: $I = \dfrac{20 \text{ V}}{800 \text{ }\Omega}$

Answer: $I = 25$ mA

8. Formula: $X_L = 2\pi f L$

Substitution: $X_L = 2 \times \pi \times 400$ Hz $\times 300$ mH

Answer: $X_L = 754 \text{ }\Omega$

9. Formula: $L = \dfrac{X_L}{2\pi f}$

Substitution: $L = \dfrac{200 \text{ }\Omega}{2 \times \pi \times 40 \text{ Hz}}$

Answer: $L = 0.796$ H

10. Formula: $f = \dfrac{X_L}{2\pi L}$

Substitution: $f = \dfrac{3.14 \text{ k}\Omega}{2 \times \pi \times 1 \text{ H}}$

Answer: $f = 500$ Hz

11. An increase in f causes an increase in X_L.

12. A decrease in L causes a decrease in X_L.

13. Formula: $X_{L_T} = X_{L_1} + X_{L_2} + X_{L_3}$

Substitution: $X_{L_T} = 40 \text{ }\Omega + 30 \text{ }\Omega + 80 \text{ }\Omega$

Answer: $X_{L_T} = 150 \text{ }\Omega$

14. Formula: $\dfrac{1}{X_{L_T}} = \dfrac{1}{X_{L_1}} + \dfrac{1}{X_{L_2}} + \dfrac{1}{X_{L_3}}$

Substitution: $\dfrac{1}{X_{L_T}} = \dfrac{1}{40 \text{ }\Omega} + \dfrac{1}{80 \text{ }\Omega} + \dfrac{1}{60 \text{ }\Omega}$

Answer: $X_{L_T} = 18.5 \text{ }\Omega$

15. V_L

$\theta = 90°$

I

16. Formula: $X_L = 2\pi f L$

Substitution: $X_L = 2 \times \pi \times 1500$ Hz $\times 42.5$ mH

Answer: $X_L = 401 \text{ }\Omega$

17. Formula: $Z = \sqrt{X_L^2 \times R^2}$

Substitution: $Z = \sqrt{(401 \text{ }\Omega)^2 + (300 \text{ }\Omega)^2}$

Answer: $Z = 501 \text{ }\Omega$

18.

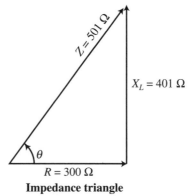

$Z = 501 \text{ }\Omega$ $X_L = 401 \text{ }\Omega$ θ $R = 300 \text{ }\Omega$

Impedance triangle

19. Formula: $\theta = \tan^{-1} \dfrac{X_L}{R}$

Substitution: $\theta = \tan^{-1} \dfrac{401 \text{ }\Omega}{300 \text{ }\Omega}$

Answer: $\theta = 53.2°$

20. Formula: $I = \dfrac{V}{Z}$

Substitution: $I = \dfrac{30 \text{ V}}{501 \text{ }\Omega}$

Answer: $I = 60$ mA

21. Formula: $V = I \times R$

Substitution: $V_R = 60$ mA $\times 300 \text{ }\Omega$

Answer: $V_R = 18$ V

Substitution: $V_L = 60$ mA $\times 401 \text{ }\Omega$

Answer: $V_L = 24.1$ V

22.

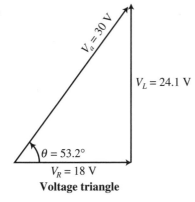

$V_a = 30$ V

$V_L = 24.1$ V

$\theta = 53.2°$

$V_R = 18$ V

Voltage triangle

24.

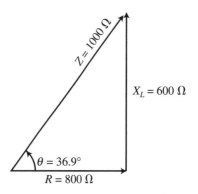

$Z = 1000\ \Omega$

$X_L = 600\ \Omega$

$\theta = 36.9°$

$R = 800\ \Omega$

23.   Formula: $Z = \sqrt{X_L^2 \times R^2}$

Substitution: $Z = \sqrt{(600\ \Omega)^2 + (800\ \Omega)^2}$

Answer: $Z = 1000\ \Omega$

25.   Formula: $\theta = \tan^{-1} \dfrac{X_L}{R}$

Substitution: $\theta = \tan^{-1} \dfrac{600\ \Omega}{800\ \Omega}$

Answer: $\theta = 36.9°$

Chapter Test
18

Name: _____

Date: _____

Class: _____

Inductive Reactance and Impedance

Define the following technical terms.

1. Impedance: _____

2. Inductive reactance: _____

3. Phase angle: _____

4. Phasor: _____

5. Pythagorean theorem: _____

With each problem, write the formula, substitution, and answer.

6. Use Ohm's law to calculate the inductive reactance of a circuit with an applied voltage of 100 volts and current through the inductor of 2 amps.

Formula: _____

Substitution: _____

Answer: _____

7. What is the current in a circuit with X_L of 800 Ω and applied voltage of 20 volts?

Formula: _____

Substitution: _____

Answer: _____

8. Use the inductive reactance formula to calculate X_L of a 300 mH inductor in a circuit with a frequency of 400 hertz.

Formula: _____

Substitution: _____

Answer: _____

9. If the inductive reactance in a circuit is 200 ohms, calculate the value of inductance if the frequency is 40 hertz.

Formula: _____

Substitution: _____

Answer: _____

10. If the inductance in a circuit is 1 H, what frequency would be required to produce a reactance of 3.14 kilohms?

Formula: _____

Substitution: _____

Answer: _____

11. Using the inductive reactance formula as a guide, what is the effect on X_L when there is an increase in frequency?

_____

12. Using the inductive reactance formula as a guide, what is the effect on X_L when there is a decrease in the value of

inductance? _____

_____

13. Calculate the total reactance of three inductive reactances connected in series: $X_{L_1} = 40\ \Omega$, $X_{L_2} = 30\ \Omega$, $X_{L_3} = 80\ \Omega$.

Formula: _____

Substitution: _____

Answer: _____

14. Calculate the total reactance of three inductive reactances connected in parallel: $X_{L_1} = 40\ \Omega$, $X_{L_2} = 80\ \Omega$, $X_{L_3} = 60\ \Omega$.

Formula: _____

Substitution: _____

Answer: _____

15. Use the space below to draw a phasor diagram showing the relationship between voltage and current in a purely inductive circuit.

For questions 16 though 22, calculate the requested values for a series ac circuit with the values: R = 300 Ω, V = 30 V, f = 1500 Hz, L = 42.5 mH.

16. Inductive reactance, X_L:

Formula: _____

Substitution: _____

Answer: _____

17. Impedance, Z:

Formula: _____

Substitution: _____

Answer: _____

18. Plot the impedance triangle.

Name: ______________________________

19. Operating angle θ:

Formula: ______________________________

Substitution: ______________________________

Answer: ______________________________

20. Circuit current:

Formula: ______________________________

Substitution: ______________________________

Answer: ______________________________

21. Voltage drops for R and X_L:

Formula: _____

Substitution for V_R: _____

Answer for V_R: _____

Substitution for V_L: _____

Answer for V_L: _____

22. Draw the voltage triangle.

For questions 23 through 25, calculate the requested values for a series ac circuit with the values: R = 800 Ω, X_L = 600 Ω, V = 20 V.

23. Impedance, Z:

Formula: _____

Substitution: _____

Answer: _____

24. Plot the impedance triangle.

25. Operating angle θ:

Formula: _____

Substitution: _____

Answer: _____

Capacitive Reactance and Impedance

OBJECTIVES

After studying this chapter, students should be able to:

- Define capacitive reactance and describe its effect on an ac circuit.
- Use Ohm's law to calculate capacitive reactance with current and voltage measurements.
- Use the capacitive reactance formula with values of frequency and capacitance.
- Calculate total reactance in series and parallel capacitive reactance circuits.
- Identify the relationships of voltage and current in a purely capacitive circuit.
- Calculate voltage drops and impedance in a series RC ac circuit.
- Plot the phasor diagrams for voltage and impedance in a series circuit.
- Calculate phase shift and opposite angle for a series circuit.
- Describe the changes in circuit parameters for series circuits with a value of resistance larger and smaller than the capacitive reactance.
- Calculate branch currents and total current in a parallel RC ac circuit.
- Calculate impedance in a parallel circuit.
- Plot the current triangle for a parallel RC circuit.
- Compare the effects of a resistance larger and smaller than the value of capacitive reactance in a parallel circuit.
- Read an oscilloscope screen to measure phase shift and opposite angle in an RC circuit.

INSTRUCTIONAL MATERIALS

Text: Pages 565–602
  Test Your Knowledge Questions, Pages 600–602
Study Guide: Pages 159–170
Laboratory Manual: Pages 203–221

ANSWERS TO TEXTBOOK

Test Your Knowledge, Pages 600–602

1. Capacitive reactance is the ac resistance of a capacitor. It limits the current in an ac circuit.
2. $X_C = 50 \ \Omega$
3. $X_C = 63.7 \ \Omega$
4. $X_T = 250 \ \Omega$
5. $X_T = 40 \ \Omega$
6. Voltage lags current by 90°.
7. $X_C = 500 \ \Omega$
   $Z = 707 \ \Omega$
   $I = 70.7 \ \text{mA}$
   $V_R = 35.4 \ \text{V}$
   $V_C = 35.4 \ \text{V}$
8. Refer to figure 19-16.
9. Refer to figure 19-16.
10. $\theta = -45°$
    $\phi = -45°$
11. $X_C = 50 \ \Omega$
    $I_R = 1 \ \text{A}$
    $I_C = 0.5 \ \text{A}$
    $I_T = 1.12 \ \text{A}$
    $Z = 22.3 \ \Omega$
12. Refer to figure 19-21.
13. $\theta = 26.6°$
    $\phi = 63.4°$
14. Refer to figures 19-17 to 19-19 (series circuits).
15. When $R = X_C$, the phase shift is 45°.
    When R is greater than X_C, the phase shift is less than 45°.
    When R is less than X_C, the phase shift is greater than 45°.
16. Impedance has characteristics of the larger value.
17. Refer to figures 19-22 to 19-24 (parallel circuits).
18. When $R = X_C$, the phase shift is 45°.
    When R is greater than X_C, current is larger in X_C, therefore, the phase shift is greater than 45°.
    When R is less than X_C, current is smaller in X_C, therefore, the phase shift is less than 45°.

19. Impedance has characteristics of the larger value.
20. $\theta = -43.2°$
    $\phi = 46.8°$

ANSWERS TO STUDY GUIDE

Pages 159–170

1. d.
2. c.

3. Formula: $X_C = \dfrac{V_a}{I}$

   Substitution: $X_C = \dfrac{40 \text{ V}}{250 \text{ mA}}$

   Answer: $X_C = 160 \ \Omega$

4. Formula: $I = \dfrac{V_a}{X_C}$

   Substitution: $I = \dfrac{15 \text{ V}}{5 \text{ k}\Omega}$

   Answer: $I = 3 \text{ mA}$

5. Formula: $X_C = \dfrac{1}{2\pi f C}$

   Substitution: $X_C = \dfrac{1}{2 \times \pi \times 2 \text{ kHz} \times 5 \ \mu\text{F}}$

   Answer: $X_C = 15.9 \ \Omega$

6. Formula: $C = \dfrac{1}{2\pi f X_C}$

   Substitution: $C = \dfrac{1}{2 \times \pi \times 50 \text{ Hz} \times 130 \ \Omega}$

   Answer: $C = 24.5 \ \mu\text{F}$

7. Formula: $f = \dfrac{1}{2\pi C X_C}$

   Substitution: $f = \dfrac{1}{2 \times \pi \times 0.2 \ \mu\text{F} \times 5 \text{ k}\Omega}$

   Answer: $f = 159 \text{ Hz}$

8. An increase in frequency will decrease X_C.
9. A decrease in capacitance will increase X_C.

10. Formula: $X_{C_T} = X_{C_1} + X_{C_2} + X_{C_3} + ... X_{C_N}$

    Substitution: $X_{C_T} = 100 \ \Omega + 200 \ \Omega + 300 \ \Omega$

    Answer: $X_{C_T} = 600 \ \Omega$

11. Formula: $\dfrac{1}{X_{C_T}} = \dfrac{1}{X_{C_1}} + \dfrac{1}{X_{C_2}} + \dfrac{1}{X_{C_3}} + ... \dfrac{1}{X_{C_N}}$

    Substitution: $\dfrac{1}{X_{C_T}} = \dfrac{1}{50 \ \Omega} + \dfrac{1}{200 \ \Omega} + \dfrac{1}{100 \ \Omega}$

    Answer: $X_{C_T} = 28.6 \ \Omega$

12. Voltage lags current by 90°. See textbook figure 19-14.

13. a. Formula: $X_C = \dfrac{1}{2\pi f C}$

    Substitution: $X_C = \dfrac{1}{2 \times \pi \times 50 \text{ Hz} \times 5.3 \ \mu\text{F}}$

    Answer: $X_C = 600 \ \Omega$

   b. Formula: $Z = \sqrt{X_C^2 + R^2}$

    Substitution: $Z = \sqrt{(600)^2 + (600)^2}$

    Answer: $Z = 849 \ \Omega$

   c. Student sketch of impedance triangle.

   d. Formula: $\theta = \tan^{-1} \dfrac{-X_C}{R}$

    Substitution: $\theta = \tan^{-1} \dfrac{-600 \ \Omega}{600 \ \Omega}$

    Answer: $\theta = -45°$

   e. Formula: $\phi = -90° - \theta$
    Substitution: $\phi = -90° + 45°$
    Answer: $\phi = -45°$

   f. Formula: $I = V/Z$
    Substitution: $I = 10 \text{ V}/849 \ \Omega$
    Answer: $I = 11.8 \text{ mA}$

   g. Formula: $V_R = I \times R$ and $V_C = I \times X_C$
    Substitution for V_R: $V_R = 11.8 \text{ mA} \times 600 \ \Omega$
    Answer: $V_R = 7.08 \text{ V}$
    Substitution for V_C: $V_C = 11.8 \text{ mA} \times 600 \ \Omega$
    Answer: $V_L = 7.08 \text{ V}$

   h. Student sketch of voltage triangle.

14. a. $Z = 22.4 \ \Omega$
    b. Student sketch of impedance triangle.
    c. $\theta = -26.6°$
    d. $\phi = -63.4°$
    e. $I = 1.34 \ A$
    f. $V_R = 26.8 \ V$
       $V_C = 13.4 \ V$
    g. Student sketch of voltage triangle.

15. a. $Z = 224 \ \Omega$
    b. Student sketch of impedance triangle.
    c. $\theta = -63.4°$
    d. $\phi = -26.6°$
    e. $I = 0.268 \ A$
    f. $V_R = 26.8 \ V$
       $V_C = 53.6 \ V$
    g. Student sketch of voltage triangle.

16. a. Formula: $X_C = \dfrac{1}{2\pi f C}$

    Substitution: $X_C = \dfrac{1}{2 \times \pi \times 400 \ Hz \times 0.5 \ \mu F}$

    Answer: $X_C = 796 \ \Omega$ (approximately 800 Ω)

    b. Formula: $I_R = V/R$
       Substitution: $I_R = 20 \ V/800 \ \Omega$
       Answer: $I_R = 25 \ mA$

    c. Formula: $I_C = V/X_C$
       Substitution: $I_C = 20 \ V/800 \ \Omega$
       Answer: $I_C = 25 \ mA$

    d. Formula: $I_T = \sqrt{I_L^2 + I_R^2}$
       Substitution: $I_T = \sqrt{(25 \ mA)^2 + (25 \ mA)^2}$
       Answer: $I_T = 35.4 \ mA$

    e. Student sketch of current triangle.

    f. Formula: $\theta = \tan^{-1} \dfrac{I_C}{I_R}$

       Substitution: $\theta = \tan^{-1} \dfrac{25 \ mA}{25 \ mA}$

       Answer: $\theta = 45°$

    g. Formula: $Z = V_a/I_T$
       Substitution: $Z = 20 \ V/35.4 \ mA$
       Answer: $Z = 565 \ \Omega$

17. a. $I_R = 20 \ mA$
    b. $I_C = 40 \ mA$
    c. $I_T = 44.7 \ mA$
    d. Student sketch of current triangle.
    e. $\theta = 63.4°$
    f. $Z = 1074 \ \Omega$

18. a. $I_R = 20 \ mA$
    b. $I_C = 10 \ mA$
    c. $I_T = 22.4 \ mA$
    d. Student sketch of current triangle.
    e. $\theta = 26.6°$
    f. $Z = 1071 \ \Omega$

19. a. $360° = 8$ divisions
    b. 1 division $= 45°$
    c. Difference between waveforms $= -1.2$ divisions
    d. $\theta = -54°$

20. a. $360° = 10$ divisions
    b. 1 division $= 36°$
    c. Difference between waveforms $= -0.8$ divisions
    d. $\theta = -28.8°$

ANSWERS TO CHAPTER TEST IN THE INSTRUCTOR'S MANUAL

Pages 235–243

1. Capacitive reactance: The ac resistance of a capacitor.

2. Impedance: The total ac resistance of a circuit containing both resistance and reactance.

3. Formula: $X_C = \dfrac{V}{I}$

   Substitution: $X_C = \dfrac{150 \ V}{750 \ mA}$

   Answer: $X_C = 200 \ \Omega$

4. Formula: $I = \dfrac{V}{X_C}$

   Substitution: $I = \dfrac{60 \ V}{20 \ k\Omega}$

   Answer: $I = 3 \ mA$

5. Formula: $X_C = \dfrac{1}{2\pi f C}$

    Substitution: $X_C = \dfrac{1}{2 \times \pi \times 4 \text{ kHz} \times 10 \text{ }\mu\text{F}}$

    Answer: $X_C = 3.98 \text{ }\Omega$

6. Formula: $C = \dfrac{1}{2\pi f X_C}$

    Substitution: $C = \dfrac{1}{2 \times \pi \times 60 \text{ Hz} \times 250 \text{ }\Omega}$

    Answer: $C = 10.6 \text{ }\mu\text{F}$

7. Formula: $f = \dfrac{1}{2\pi C X_C}$

    Substitution: $f = \dfrac{1}{2 \times \pi \times 0.02 \text{ }\mu\text{F} \times 5 \text{ k}\Omega}$

    Answer: $f = 1592 \text{ Hz}$

8. An increase in f causes a decrease in X_C.
9. A decrease in C causes an increase in X_C.

10. Formula: $X_{C_T} = X_{C_1} + X_{C_2} + X_{C_3}$
    Substitution: $X_{C_T} = 15 \text{ }\Omega + 25 \text{ }\Omega + 30 \text{ }\Omega$
    Answer: $X_{C_T} = 70 \text{ }\Omega$

11. Formula: $\dfrac{1}{X_{C_T}} = \dfrac{1}{X_{C_1}} + \dfrac{1}{X_{C_2}} + \dfrac{1}{X_{C_3}}$

    Substitution: $\dfrac{1}{X_{C_T}} = \dfrac{1}{5 \text{ }\Omega} + \dfrac{1}{20 \text{ }\Omega} + \dfrac{1}{10 \text{ }\Omega}$

    Answer: $X_{C_T} = 2.86 \text{ }\Omega$

12.

    $\theta = -90°$

    I_C

    V_C

13. Formula: $X_C = \dfrac{1}{2\pi f C}$

    Substitution: $X_C = \dfrac{1}{2 \times \pi \times 120 \text{ Hz} \times 1.5 \text{ }\mu\text{F}}$

    Answer: $X_C = 884 \text{ }\Omega$

14. Formula: $Z = \sqrt{X_C^2 \times R^2}$
    Substitution: $Z = \sqrt{(884 \text{ }\Omega)^2 + (1200 \text{ }\Omega)^2}$
    Answer: $Z = 1490 \text{ }\Omega$

15.

    $R = 1200 \text{ }\Omega$

    $Z = 1490 \text{ }\Omega$

    $X_C = 884 \text{ }\Omega$

16. Formula: $\theta = \tan^{-1} \dfrac{-X_C}{R}$

    Substitution: $\theta = \tan^{-1} \dfrac{-884 \text{ }\Omega}{1200 \text{ }\Omega}$

    Answer: $\theta = -36.4°$

17. Formula: $I = \dfrac{V}{Z}$

    Substitution: $I = \dfrac{40 \text{ V}}{1490 \text{ }\Omega}$

    Answer: $I = 26.8 \text{ mA}$

18. Formula: $V_R = I \times R$
    Substitution: $V_R = 26.8 \text{ mA} \times 1200 \text{ }\Omega$
    Answer: $V_R = 32.2 \text{ V}$
19. Formula: $V_C = I \times X_C$
    Substitution: $V_C = 26.8 \text{ mA} \times 884 \text{ }\Omega$
    Answer: $V_C = 23.7 \text{ V}$

20.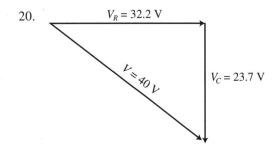

    $V_R = 32.2 \text{ V}$

    $V = 40 \text{ V}$

    $V_C = 23.7 \text{ V}$

Chapter Test

19

Capacitive Reactance and Impedance

Define the following technical terms.

1. Capacitive reactance: _____

   _____

2. Impedance: _____

   _____

With each problem, write the formula, substitution, and answer.

3. Use Ohm's law to calculate capacitive reactance in a series circuit with 750 mA current and applied voltage of 150 V.

   Formula: _____

   Substitution: _____

   Answer: _____

4. What is the current in a circuit with X_C of 20 kΩ and applied voltage of 60 volts?

Formula: _____

Substitution: _____

Answer: _____

5. Use the capacitive reactance formula to calculate X_C in a circuit with a frequency of 4 kHz and 10 μF capacitance.

Formula: _____

Substitution: _____

Answer: _____

6. The capacitive reactance in a circuit is 250 ohms. Calculate the value of capacitance if the frequency is 60 hertz.

Formula: _____

Substitution: _____

Answer: _____

7. If the capacitance in a circuit is 0.02 μF, what frequency would be required to produce a reactance of 5 kilohms?

Formula: _____

Substitution: _____

Answer: _____

8. Using the capacitive reactance formula as a guide, what is the effect on X_C when there is an increase in frequency?

   _____

9. Using the capacitive reactance formula as a guide, what is the effect on X_C when there is a decrease in the value of

   capacitance? _____

   _____

10. Calculate the total reactance when three capacitive reactances are connected in series. The individual values are:
    $X_{C_1} = 15\ \Omega$, $X_{C_2} = 25\ \Omega$, $X_{C_3} = 30\ \Omega$.

    Formula: _____

    Substitution: _____

    Answer: _____

11. Calculate the total reactance when three capacitive reactances are connected in parallel. The individual values are: $X_{C_1} = 5\ \Omega$, $X_{C_2} = 20\ \Omega$, $X_{C_3} = 10\ \Omega$.

Formula: _____

Substitution: _____

Answer: _____

12. Use the space below to draw a phasor diagram showing the relationship between voltage and current in a purely capacitive circuit.

For questions 13 through 20, calculate the requested values for a series ac circuit with these values: R = 1200 Ω, V = 40 V, f = 120 Hz, C = 1.5 μF.

13. Capacitive reactance, X_C:

Formula: _____

Substitution: _____

Answer: _____

14. Impedance, Z:

Formula: _____

Substitution: _____

Answer: _____

15. Plot the impedance triangle.

16. Operating angle θ:

Formula: _____

Substitution: _____

Answer: _____

17. Circuit current:

Formula: _____

Substitution: _____

Answer: _____

18. Voltage drop for *R*:

Formula: _____

Substitution: _____

Answer: _____

19. Voltage drop X_C:

Formula: _____

Substitution: _____

Answer: _____

20. Draw the voltage triangle.

Phasors and Complex Numbers

OBJECTIVES

After studying this chapter, students should be able to:

- Use phasor addition to solve series and parallel circuits.
- Use phasor addition to determine the effective reactance in a circuit with both inductance and capacitance.
- Mathematically represent the impedance of a circuit as resistance +/–j-operator for the value of reactance (retangular form).
- Mathematically represent the impedance of a circuit as a quantity at its phase angle (polar form).
- Convert complex numbers from rectangular to polar and polar to rectangular.
- Write complex numbers using the values given in a circuit.
- Use complex numbers to solve series and parallel circuits.

INSTRUCTIONAL MATERIALS

Text: Pages 603–640
Test Your Knowledge Questions, Pages 638–640
Study Guide: Pages 171–188
Laboratory Manual: Pages 223–235

ANSWERS TO TEXTBOOK

Test Your Knowledge, Pages 638–640

1. Resistance and reactance.
2. Student sketches of phasor diagrams.
3. 20 Ω inductive
4. X_{net} = 300 Ω capacitive
   Z = 500 Ω
   Student phasor diagram.

5. X_{net} = 10 Ω capacitive
   Z = 26.9 Ω
   I = 1.86 A
   θ = –21.8°
   V_R = 46.5 V
   V_C = 186 V
   V_L = 167 V
   Student phasor diagram.
6. R_T = 100 Ω, X_{L_T} = 200 Ω, X_{C_T} = 100 Ω
   X_{net} = 100 Ω inductive
   Z = 141 Ω
   I = 0.707 A
   θ = 45°
   Student phasor diagram.
7. I_C = 0.25 A
   I_L = 0.167 A
   I_{circ} = 0.167 A
   I_{net} = 0.083 A capacitive
   X_{net} = 120 Ω capacitive
   Student phasor diagram.
8. I_R = 0.25 A
   I_C = 0.5 A
   I_L = 0.125 A
   I_{net} = 0.375 A
   I_T = 0.451 A
   Z = 55.4 Ω

   Student phasor diagram.
9. a. Rectangular: Z = 40 + j50 Ω
      Polar: Z = 64 Ω $\angle 51.3°$
   b. Rectangular: Z = 5 – j3 kΩ
      Polar: Z = 5.83 kΩ $\angle –31°$
   c. Rectangular: Z = 0 + j35 Ω
      Polar: Z = 35 Ω $\angle 90°$
   d. Rectangular: Z = 100 Ω + j0
      Polar: Z = 100 Ω $\angle 0°$
   e. Rectangular: Z = 80 + j60 Ω
      Polar: Z = 100 Ω $\angle 36.9°$
   f. Rectangular: Z = 20 – j10 Ω
      Polar: Z = 22.4 Ω $\angle –26.6°$

g. Rectangular: $Z = 300 + j400\ \Omega$
    Polar: $Z = 500\ \Omega\ \angle 53.1°$
h. Rectangular: $Z = 25 - j75\ \Omega$
    Polar: $Z = 79.1\ \Omega\ \angle -71.6°$
i. Rectangular: $Z = 4 + j4\ k\Omega$
    Polar: $Z = 5.66\ k\Omega\ \angle 45°$
j. Rectangular: $Z = 10 - j10\ \Omega$
    Polar: $Z = 14.1\ \Omega\ \angle -45°$

Student phasor diagrams for a through j.

10. a. Rectangular: $7.07 + j7.07\ \Omega$
    b. Polar: $43\ \Omega\ \angle 35.5°$
    c. Rectangular: $13 - j7.5\ \Omega$
    d. Polar: $112\ \Omega\ \angle -63.4°$
    e. Rectangular: $40 + j0\ \Omega$
    f. Polar: $56.6\ \Omega\ \angle 45°$
    g. Rectangular: $79.9 - j60.2\ \Omega$
    h. Polar: $150\ \Omega\ \angle -90°$
    i. Rectangular: $0 + j50\ \Omega$
    j. Polar: $10\ \Omega\ \angle 0°$

11. a. Impedance:
        Rectangular: $Z = 50 - j35\ \Omega$
        Polar: $Z = 61\ \Omega\ \angle -35°$
    b. Current: $I = 0.41\ A\ \angle 35°$ (polar)
    c. Resistor voltage drop:
        Polar: $V_R = 20.5\ V\ \angle 35°$
        Rectangular: $V_R = 16.8 + j11.8\ V$
    d. Capacitive reactance voltage drop:
        Polar: $V_C = 14.4\ V\ \angle -55°$
        Rectangular: $V_C = 8.26 - j11.8\ V$

12. a. Inductive current:
        Polar: $I_L = 0.5\ A\ \angle -90°$
        Rectangular: $I_L = 0 - j0.5\ A$
    b. Resistive current:
        Polar: $I_R = 1\ A\ \angle 0°$
        Rectangular: $I_R = 1 + j0\ A$
    c. Total current:
        Rectangular: $I_T = 1 - j0.5\ A$
        Polar: $I_T = 1.12\ A\ \angle -26.6°$
    d. Total impedance:
        Polar: $Z = 89.3\ \Omega\ \angle 26.6°$
        Rectangular: $Z = 79.8 + j40\ \Omega$

13. a. Impedance of branch A:
        Rectangular: $Z_A = 50 - j50\ \Omega$
        Polar: $Z_A = 70.7\ \Omega\ \angle -45°$
        Impedance of branch B:
        Rectangular: $Z_B = 100 + j150\ \Omega$
        Polar: $Z_B = 180\ \Omega\ \angle 56.3°$
        Impedance of branch C:
        Rectangular: $Z_C = 200 + j0$
        Polar: $Z_C = 200\ \Omega\ \angle 0°$

b. Student schematic of equivalent circuit.
c. Current in branch A:
    Polar: $I_A = 1.41\ A\ \angle 45°$
    Rectangular: $I_A = 1 + j1\ A$
    Current in branch B:
    Polar: $I_B = 0.556\ A\ \angle -56.3°$
    Rectangular: $I_B = 0.308 - j0.463\ A$
    Current in branch B:
    Polar: $I_C = 0.5\ \angle 0°$
    Rectangular: $I_C = 0.5 + j0$
d. Total current:
    Rectangular: $I_T = 1.808 + j0.537\ A$
    Polar: $I_T = 1.89\ A\ \angle 16.5°$
e. Total impedance:
    Polar: $Z_T = 52.9\ \Omega\ \angle -16.5°$
    Rectangular: $Z_T = 50.7 - j15\ \Omega$

14. a. Parallel impedances:
        Impedance of branch A:
        $Z_A = 30 - j40\ \Omega$ (Rectangular)
        $Z_A = 50\ \angle -53.1°$ (Polar)
        Impedance of branch B:
        $Z_B = 50 + j20\ \Omega$ (Rectangular)
        $Z_B = 53.9\ \angle 21.8°$ (Polar)
        Equivalent impedance of parallel branches:
        $Z_{A,B} = 31.2 - j9.7\ \Omega$ (Rectangular)
    b. Student schematic of equivalent series circuit.
    c. Total impedance:
        $Z_T = 41.2 - j19.7\ \Omega$ (Rectangular)

ANSWERS TO STUDY GUIDE

Pages 171–188

1. e.
2. h.
3. d.
4. b.
5. c.
6. i.
7. f.
8. g.
9. a.
10. j.
11. Student graph of phasors.
12. a. Net reactance:
        Formula: $X_{net} = X_C - X_L$
        Substitution: $X_{net} = 250\ \Omega - 200\ \Omega$
        Answer: $X_{net} = 50\ \Omega$
    b. Capacitive.

13. a. Net reactance:
     Formula: $X_{net} = X_L - X_C$
     Substitution: $X_{net} = 170\ \Omega - 50\ \Omega$
     Answer: $X_{net} = 120\ \Omega$ (inductive)

  b. Impedance:
     Formula: $Z = \sqrt{R_T^2 + X_{net}^2}$
     Substitution: $Z = \sqrt{(60\ \Omega)^2 + (120\ \Omega)^2}$
     Answer: $Z = 134\ \Omega$

14. a. Net reactance:
     Formula: $X_{net} = X_C - X_L$
     Substitution: $X_{net} = 600\ \Omega - 350\ \Omega$
     Answer: $X_{net} = 250\ \Omega$ (capacitive)

  b. Impedance:
     Formula: $Z = \sqrt{R_T^2 + X_{net}^2}$
     Substitution: $Z = \sqrt{(250\ \Omega)^2 + (250\ \Omega)^2}$
     Answer: $Z = 354\ \Omega$

  c. Current:
     Formula: $I = V/Z$
     Substitution: $I = 100\ \text{V} \div 354\ \Omega$
     Answer: $I = 0.282$ A (polar)

  d. Phase angle:

     Formula: $\theta = \tan^{-1} \dfrac{X_{net}}{R}$

     Substitution: $\theta = \tan^{-1} \dfrac{250\ \Omega}{250\ \Omega}$

     Answer: $\theta = \angle 45°$

  e. Voltage drop across R:
     Formula: $V_R = I \times R$
     Substitution $V_R = 0.282$ A $\times 250\ \Omega$
     Answer: $V_R = 70.5$ V

  f. Voltage drop across X_C:
     Formula: $V_C = I \times X_C$
     Substitution $V_C = 0.282$ A $\times 600\ \Omega$
     Answer: $V_C = 169$ V

  h. Voltage drop across X_L:
     Formula: $V_L = I \times X_L$
     Substitution $V_L = 0.282$ A $\times 350\ \Omega$
     Answer: $V_L = 98.7$ V

15. a. Net reactance:
     Formula: $X_{net} = X_C - X_L$
     Substitution: $X_{net} = 8\ \Omega + 8\ \Omega - 4\ \Omega - 4\ \Omega$
     Answer: $X_{net} = 8\ \Omega$ (capacitive)

  b. Impedance:
     Formula: $Z = \sqrt{R_T^2 + X_{net}^2}$
     Substitution: $Z = \sqrt{(12\ \Omega)^2 + (8\ \Omega)^2}$
     Answer: $Z = 14.4\ \Omega$

  c. Current:
     Formula: $I = V/Z$
     Substitution: $I = 10\ \text{V} \div 14.4\ \Omega$
     Answer: $I = 0.694$ A

  d. Phase angle:

     Formula: $\theta = \tan^{-1} \dfrac{X_{net}}{R_T}$

     Substitution: $\theta = \tan^{-1} \dfrac{-8\ \Omega}{12\ \Omega}$

     Answer: $\theta = \angle -33.7°$

  e. Voltage drop across R_T:
     Formula: $V_R = I \times R_T$
     Substitution: $V_R = 0.694$ A $\times 12\ \Omega$
     Answer: $V_R = 8.33$ V

  f. Voltage across net reactance:
     Formula: $V_{Xnet} = I \times X_{net}$
     Substitution: $V_{Xnet} = 0.694$ A $\times 8\ \Omega$
     Answer: $V_{Xnet} = 5.55$ V

16. a. Capacitive branch current:

     Formula: $I_C = \dfrac{V}{X_C}$

     Substitution: $I_C = \dfrac{10\ \text{V}}{20\ \Omega}$

     Answer: $I_C = 0.5$ A

  b. Inductive branch current:

     Formula: $I_L = \dfrac{V}{X_L}$

     Substitution: $I_L = \dfrac{10\ \text{V}}{10\ \Omega}$

     Answer: $I_L = 1$ A

c. Circulation current:

No formula needed. Circulation current is equal to the smaller branch current.

$I_{circ} = 0.5$ A

d. Net current:

Formula: $I_{net} = I_L - I_C$

Substitution: $I_{net} = 1$ A $- 0.5$ A

Answer: $I_{net} = 0.5$ A

e. Net reactance:

Formula: $X_{net} = \dfrac{V}{I_{net}}$

Substitution: $X_{net} = \dfrac{10\text{ V}}{0.5\text{ A}}$

Answer: $X_{net} = 20$ Ω

17. a. Resistance branch current:

Formula: $I_R = \dfrac{V}{R}$

Substitution: $I_R = \dfrac{20\text{ V}}{1\text{ k}\Omega}$

Answer: $I_R = 20$ mA

b. Capacitance branch current:

Formula: $I_C = \dfrac{V}{X_C}$

Substitution: $I_C = \dfrac{20\text{ V}}{2\text{ k}\Omega}$

Answer: $I_C = 10$ mA

c. Inductance branch current:

Formula: $I_L = \dfrac{V}{X_L}$

Substitution: $I_L = \dfrac{20\text{ V}}{4\text{ k}\Omega}$

Answer: $I_L = 5$ mA

d. Circulation current:

Answer: $I_{circ} = 5$ mA

e. Net reactive current:

Formula: $I_{net} = I_C - I_L$

Substitution: $I_{net} = 10$ mA $- 5$ mA

Answer: $I_{net} = 5$ mA

f. Total current:

Formula: $I_T = \sqrt{I_R^2 + I_{net}^2}$

Substitution: $I_T = \sqrt{(20\text{ mA})^2 + (5\text{ mA})^2}$

Answer: $I_T = 20.6$ mA

g. Impedance:

Formula: $Z = \dfrac{V}{I_T}$

Substitution: $Z = \dfrac{20\text{ V}}{20.6\text{ mA}}$

Answer: $Z = 971$ Ω

18. a. Rectangular: $Z = 50 + j75$ Ω
Polar: $Z = 90.1$ Ω $\angle 56.3°$
b. Rectangular: $Z = 3 - j2$ kΩ
Polar: $Z = 3.6$ kΩ $\angle -33.7°$
c. Rectangular: $Z = 60 + j0$ Ω
Polar: $Z = 60$ Ω $\angle 0°$
d. Rectangular: $Z = 500 - j800$ Ω
Polar: $Z = 943$ Ω $\angle -58°$
e. Rectangular: $Z = 75 + j60$ Ω
Polar: $Z = 96$ Ω $\angle 38.7°$
f. Rectangular: $Z = 0 - j150$ Ω
Polar: $Z = 150$ Ω $\angle -90°$
g. Rectangular: $Z = 35 + j45$ Ω
Polar: $Z = 57$ Ω $\angle 52.1°$
h. Rectangular: $Z = 20 - j90$ kΩ
Polar: $Z = 92.2$ Ω $\angle -77.5°$
i. Rectangular: $Z = 24 + j24$ Ω
Polar: $Z = 33.9$ Ω $\angle 45°$
j. Rectangular: $Z = 75 - j75$ Ω
Polar: $Z = 106$ Ω $\angle -45°$

19. a. Rectangular: $17.7 + j17.7$ Ω
b. Polar: 90.1 Ω $\angle 33.7°$
c. Rectangular: $156 - j90$ Ω
d. Polar: 14.4 Ω $\angle -33.7°$
e. Rectangular: $30.1 + j39.9$ Ω
f. Polar: 14.1 Ω $\angle 45°$
g. Rectangular: $160 - j120$ Ω
h. Polar: 492 Ω $\angle -66°$
i. Rectangular: $50 + j0$ Ω
j. Polar: 103 Ω $\angle 14°$

20. a. Impedance:
Rectangular: $Z = 25 - j50$ Ω
Polar: $Z = 55.9$ Ω $\angle -63.4°$

b. Current:
Formula: $I = V/Z$
Substitution: $I = 50$ V $\angle 0° \div 55.9$ Ω $\angle -63.4°$
Answer: $I = 0.894$ A $\angle 63.4°$

c. Resistance voltage drop:
Formula: $V_R = I \times R$
Substitution:
$V_R = 0.894$ A $\angle 63.4° \times 25$ Ω $\angle 0°$
Answer: $V_R = 22.4$ V $\angle 63.4°$

d. Capacitance voltage drop:
Formula: $V_C = I \times X_C$
Substitution:
$V_C = 0.894$ A $\angle 63.4° \times 50$ Ω $\angle -90°$
Answer: $V_C = 44.7$ V $\angle -26.6°$

21. a. Complex number for resistance branch:
Rectangular: $100 + j0$ Ω
Polar: 100 Ω $\angle 0°$

b. Complex number for inductance branch:
Rectangular: $0 + j200$ Ω
Polar: 200 Ω $\angle 90°$

c. Resistance branch current:
Formula: $I_R = V/R$
Substitution: $I_R = 25$ V $\angle 0° \div 100$ Ω $\angle 0°$
Answer: $I_R = 0.25$ A $\angle 0°$
Polar: $I_R = 0.25$ A $\angle 0°$
Rectangular: $I_R = 0.25 + j0$ A

d. Inductance branch current:
Formula $I_L = V/X_L$
Substitution: $I_L = 25$ V $\angle 0° \div 200$ Ω $\angle 90°$
Answer: $I_L = 0.125$ A $\angle -90°$
Polar: $I_L = 0.125$ A $\angle -90°$
Rectangular: $I_L = 0 - j0.125$ A

e. Total current:
Formula: $I_T = I_R + I_L$
Substitution:
$I_T = (0.25 + j0$ A$) + (0 - j0.125$ A$)$
Answer: $I_T = 0.25 - j0.125$ A
Polar: $I_T = 0.28$ A $\angle -26.6°$
Rectangular: $I_T = 0.25 - j0.125$ A

f. Total impedance:
Formula: $Z = V/I_T$
Substitution:
$Z = 25$ V $\angle 0° \div 0.28$ A $\angle -26.6°$
Answer: $Z = 89.3$ Ω $\angle 26.6°$
Polar: $Z = 89.3$ Ω $\angle 26.6°$
Rectangular: $Z = 79.8 + j40$ Ω

22. a. Impedance of branch A:
Rectangular: $Z_A = 70 + j70$ Ω
Polar: $Z_A = 99$ Ω $\angle 45°$

b. Impedance of branch B:
Rectangular: $Z_B = 50 - j150$ Ω
Polar: $Z_B = 158$ Ω $\angle -71.6°$

c. Impedance of branch C:
Rectangular: $Z_C = 100 - j100$ Ω
Polar: $Z_C = 141$ Ω $\angle -45°$

d. Student sketch of equivalent circuit.

e. Admittance of branch A:

Formula: $Y_A = \dfrac{1}{Z_A}$

Substitution: $Y_A = \dfrac{1}{99 \text{ Ω } \angle 45°}$

Answer: $Y_A = 10.1$ mS $\angle -45°$

Polar: $Y_A = 10.1$ mS $\angle -45°$

Rectangular: $Y_A = 7.14 - j7.14$ mS

f. Admittance of branch B:

Formula: $Y_B = \dfrac{1}{Z_B}$

Substitution: $Y_B = \dfrac{1}{158 \text{ Ω } \angle -71.6°}$

Answer: $Y_B = 6.33$ mS $\angle 71.6°$

Polar: $Y_B = 6.33$ mS $\angle 71.6°$

Rectangular: $Y_B = 2 + j6$ mS

g. Admittance of branch C:

Formula: $Y_C = \dfrac{1}{Z_C}$

Substitution: $Y_C = \dfrac{1}{141 \text{ Ω } \angle -45°}$

Answer: $Y_C = 7.09$ mS $\angle 45°$

Polar: $Y_C = 7.09$ mS $\angle 45°$

Rectangular: $Y_C = 5.01 + j5.01$ mS

h. Total admittance:
    Formula: $Y_T = Y_A + Y_B + Y_C$
    Substitution: $Y_T = (7.14 - j7.14 \text{ mS}) +$
    $(2 + j6 \text{ mS}) + (5.01 + j5.01 \text{ mS})$
    Answer: $Y_T = 14.15 + j3.87 \text{ mS}$
    Rectangular: $Y_T = 14.15 + j3.87 \text{ mS}$
    Polar: $Y_T = 14.7 \text{ mS} \angle 15.3°$

i. Total impedance:

    Formula: $Z_T = \dfrac{1}{Y_T}$

    Substitution: $Z_T = \dfrac{1}{14.7 \text{ mS} \angle 15.3°}$

    Answer: $Z_T = 68 \text{ } \Omega \angle -15.3°$

    Polar: $Z_T = 68 \text{ } \Omega \angle -15.3°$

    Rectangular: $Z_T = 65.6 - j17.9 \text{ } \Omega$

j. Total current:
    Formula: $I_T = V/Z_T$
    Substitution:
    $I_T = 100 \text{ V} \angle 0° \div 68 \text{ } \Omega \angle -15.3°$
    Polar: $I_T = 1.47 \angle 15.3°$ A
    Rectangular: $I_T = 1.42 - j0.388$ A

23. a. Impedance of branch A:
        Rectangular: $Z_A = 40 - j30 \text{ } \Omega$
        Polar: $Z_A = 50 \text{ } \Omega \angle -36.9°$
    b. Impedance of branch B:
        Rectangular: $Z_B = 30 + j40 \text{ } \Omega$
        Polar: $Z_B = 50 \text{ } \Omega \angle 53.1°$
    c. Impedance of branch C:
        Rectangular: $Z_C = 50 - j50 \text{ } \Omega$
        Polar: $Z_C = 70.7 \text{ } \Omega \angle -45°$
    d. Student sketch of equivalent circuit.

    e. Admittance of branch A:

        Formula: $Y_A = \dfrac{1}{Z_A}$

        Substitution: $Y_A = \dfrac{1}{50 \text{ } \Omega \angle -36.9°}$

        Answer: $Y_A = 20 \text{ mS} \angle 36.9°$

        Polar: $Y_A = 20 \text{ mS} \angle 36.9°$

        Rectangular: $Y_A = 16 + j12 \text{ mS}$

f. Admittance of branch B:

    Formula: $Y_B = \dfrac{1}{Z_B}$

    Substitution: $Y_B = \dfrac{1}{50 \text{ } \Omega \angle 53.1°}$

    Answer: $Y_B = 20 \text{ mS} \angle -53.1°$

    Polar: $Y_B = 20 \text{ mS} \angle -53.1°$

    Rectangular: $Y_B = 12 - j16 \text{ mS}$

g. Admittance of branch C:

    Formula: $Y_C = \dfrac{1}{Z_C}$

    Substitution: $Y_C = \dfrac{1}{70.7 \text{ } \Omega \angle -45°}$

    Answer: $Y_C = 14.1 \text{ mS} \angle 45°$

    Polar: $Y_C = 14.1 \text{ mS} \angle 45°$

    Rectangular: $Y_C = 9.97 + j9.97 \text{ mS}$

h. Total admittance of the three parallel branches:
    Formula: $Y_T = Y_A + Y_B + Y_C$
    Substitution: $Y_T = (16 + j12 \text{ mS}) +$
    $(12 - j16 \text{ mS}) + (9.97 + j9.97 \text{ mS})$
    Answer: $Y_T = 38 + j5.97 \text{ mS}$
    Polar: $Y_T = 38.5 \text{ mS} \angle 8.9°$

i. Equivalent impedance of the three parallel branches:

    Formula: $Z_{PAR} = \dfrac{1}{Y_T}$

    Substitution: $Z_{PAR} = \dfrac{1}{38.5 \text{ mS} \angle 8.9°}$

    Answer: $Z_{PAR} = 26 \text{ } \Omega \angle -8.9°$

    Rectangular: $Z_{PAR} = 25.7 - j4.02 \text{ } \Omega$

    Polar: $Z_{PAR} = 26 \text{ } \Omega \angle -8.9°$

j.   Student sketch of equivalent circuit.
k.   Total impedance:
      Rectangular: $Z_T = 45.7 + j5.98 \ \Omega$
      Polar: $Z_T = 46.1 \ \Omega \ \angle 7.45°$

ANSWERS TO CHAPTER TEST IN THE INSTRUCTOR'S MANUAL

Pages 253–257

1.   net reactance
2.   j-operator
3.   complex
4.   rectangular
5.   polar

6.

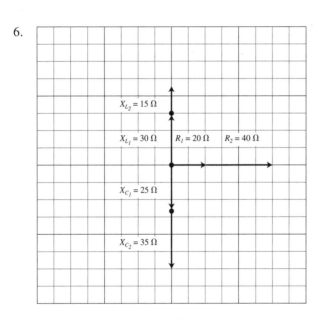

7.   a.   Formula: $X_{net} = X_L - X_C$
           Substitution: $X_{net} = 40 \ \Omega - 35 \ \Omega$
           Answer: $X_{net} = 5 \ \Omega$

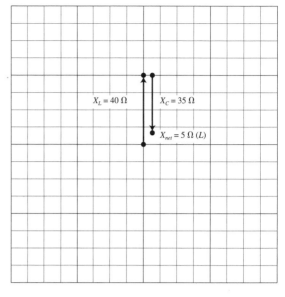

b.   Inductive

8.   a.   Formula: $X_{net} = X_L - X_C$
           Substitution: $X_{net} = 70 \ \Omega - 40 \ \Omega$
           Answer: $X_{net} = 30 \ \Omega$ (inductive)

b.   Formula: $Z = \sqrt{X_{net}^2 \times R^2}$
      Substitution: $Z = \sqrt{(30 \ \Omega)^2 + (30 \ \Omega)^2}$
      Answer: $Z = 42.4 \ \Omega$

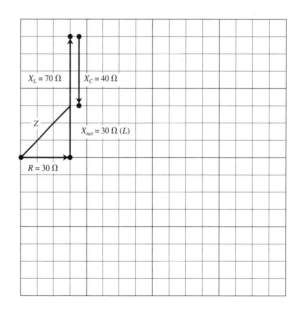

9.

	Rectangular		Polar	
	Impedance	Phasor diagram	Impedance	Phasor diagram
a. $R = 5\ \Omega$ $X_L = 8\ \Omega$	$5 + j8\ \Omega$	j8 / 5	$9.43\ \Omega\ \angle 58°$	$9.43\ \Omega\ \angle 58°$
b. $R = 60\ k\Omega$ $X_C = 40\ k\Omega$	$60 - j40\ k\Omega$	60 / j40	$72.1\ k\Omega\ \angle{-33.7°}$	$-33.7°$ $72.1\ k\Omega$
c. $R = 60\ \Omega$ $X_L = 0\ \Omega$	$60 + j0\ \Omega$	60	$60\ \Omega\ \angle 0°$	$60\ \Omega$ $0°$
d. $R = 0\ \Omega$ $X_C = 8\ \Omega$	$0 - j8\ \Omega$	$-j8$	$8\ \Omega\ \angle{-90°}$	$-90°$ $8\ \Omega$
e. $R = 60\ \Omega$ $X_L = 60\ \Omega$	$60 + j60\ \Omega$	$+j60$ / 60	$84.9\ \Omega\ \angle 45°$	$84.8\ \Omega\ \angle 45°$

10. a. Rectangular: $60 - j80\ \Omega$
       Polar: $100\ \Omega\ \angle{-53.1°}$

   b. Formula: $I = \dfrac{V}{Z}$

       Substitution: $I = \dfrac{20\ V\ \angle 0°}{100\ \Omega\ \angle{-53.1°}}$

       Answer: $I = 0.2\ A\ \angle 53.1°$

   c. Formula: $V_R = I \times R$
       Substitution: $V_R = 0.2\ A\ \angle 53.1° \times 60\ \Omega\ \angle 0°$
       Answer: $V_R = 12\ V\ \angle 53.1°$
   d. Formula: $V_C = I \times X_C$
       Substitution:
       $\qquad V_C = 0.2\ A\ \angle 53.1° \times 80\ \Omega\ \angle{-90°}$
       Answer: $V_C = 16\ V\ \angle{-36.9°}$

Chapter Test

20

Name: _____

Date: _____

Class: _____

Phasors and Complex Numbers

Fill in the blanks.

1. The effective reactance in a circuit containing both X_L and X_C is called the _____.

2. The _____ is a mathematical tool used to represent reactances or impedances as complex numbers.

3. Numbers containing a real term and an imaginary term are _____ numbers.

4. _____ coordinates are paired numbers, written in the form (real, imaginary).

5. When an impedance is expressed as a phasor with a length and direction, it is in _____ form.

6. Use the grid provided to draw the phasors for the following components: $R_1 = 20\ \Omega$, $X_{C_1} = 25\ \Omega$, $X_{L_1} = 30\ \Omega$, $R_2 = 40\ \Omega$, $X_{C_2} = 35\ \Omega$, $X_{L_2} = 15\ \Omega$.

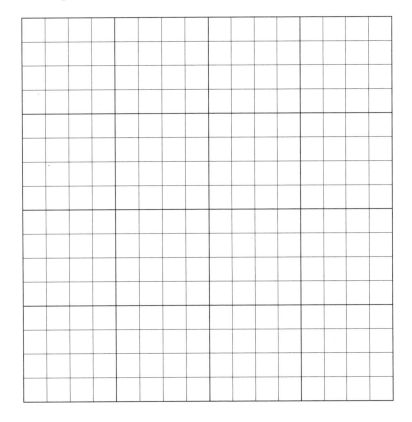

7. Use phasor addition to find the net reactance in a series circuit with $X_C = 35\ \Omega$ and $X_L = 40\ \Omega$. On the grid provided, diagram the phasor addition.

   a. Net reactance:

   Formula: _____

   Substitution: _____

   Answer: _____

b.   Is the net reactance capacitive or inductive? (circle one)

8. On the grid provided, draw an impedance triangle and solve for impedance in a series circuit with $R = 30\ \Omega$, $X_C = 40\ \Omega$, and $X_L = 70\ \Omega$. Show all necessary calculations.

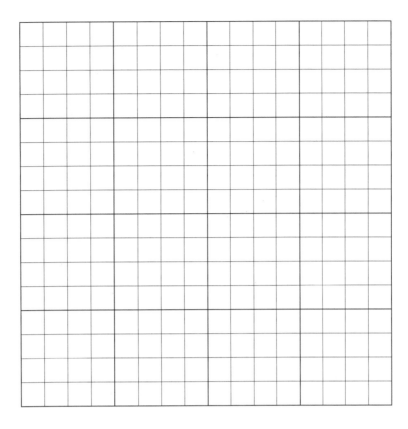

a. Net Reactance:

Formula: _____

Substitution: _____

Answer: _____

b. Impedance:

Formula: _____

Substitution: _____

Answer: _____

9. For each of the resistance and reactance combinations, find the equivalent impedance in rectangular form and polar form. Also, sketch an approximate phasor diagram to represent the impedance in each form.

	Rectangular		**Polar**	
	Impedance	Phasor diagram	Impedance	Phasor diagram
a. $R = 5\ \Omega$ $X_L = 8\ \Omega$				
b. $R = 60\ k\Omega$ $X_C = 40\ k\Omega$				
c. $R = 60\ \Omega$ $X_L = 0\ \Omega$				
d. $R = 0\ \Omega$ $X_C = 8\ \Omega$				
e. $R = 60\ \Omega$ $X_L = 60\ \Omega$				

10. Use complex numbers to find the requested values in a series circuit with an applied voltage of 20 V, $R = 60 \ \Omega$, and $X_C = 80 \ \Omega$.

    a. Impedance:

        Rectangular _____

        Polar _____

    b. Current (in polar form):

        Formula: _____

        Substitution: _____

        Answer: _____

    c. Resistance voltage drop:

        Formula: _____

        Substitution: _____

        Answer (polar): _____

    d. Capacitance voltage drop:

        Formula: _____

        Substitution: _____

        Answer (polar): _____

21

Circuit Theorems Applied to AC

OBJECTIVES

After studying this chapter, students should be able to:
*   Solve series and parallel circuits by treating each resistance and reactance combination as an impedance network.
*   Use Kirchhoff's voltage law to solve complex ac circuits.
*   Use Kirchhoff's current law to solve complex ac circuits.
*   Use the superposition theorem to solve ac circuits with two voltage sources.
*   Use Thevenin's theorem to simplify a complex ac circuit.
*   Represent delta and wye transformations as three-terminal impedance networks.

INSTRUCTIONAL MATERIALS

Text: Pages 641–666
  Test Your Knowledge Questions, Pages 662–665
Study Guide: Pages 189–202

ANSWERS TO TEXTBOOK

Test Your Knowledge, Pages 662–665

1. $Z_1 = 70.7 - j70.7 \ \Omega$ (rectangular)
    $Z_2 = 75 + j130 \ \Omega$ (rectangular)
    Answer: $Z_T = 145.7 + j59.3 \ \Omega$

2. $Y_1 = 10 \ mS \ \angle 45°$ (polar)
    $Y_1 = 7.07 + j7.07 \ mS$ (rectangular)
    $Y_2 = 6.67 \ mS \ \angle{-60°}$ (polar)
    $Y_2 = 3.33 - j5.77 \ mS$ (rectangular)
    $Y_T = 10.4 + j1.3 \ mS$ (rectangular)
    $Y_T = 10.5 \ mS \ \angle 7.13°$
    Answer: $Z_T = 95.2 \ \Omega \ \angle{-7.13°}$

3. a. $Z_1 = 43.3 + j25 \ \Omega$
    $Z_2 = 15.8 - j12.3 \ \Omega$
    $Z_3 = 28.3 - j28.3 \ \Omega$
    $Z_T = 87.4 - j15.6 \ \Omega$ (rectangular)
    $Z_T = 88.8 \ \Omega \ \angle{-10.1°}$ (polar)

    b. $I = 1.13 \ A \ \angle 10.1°$

    c. $V_{Z_1} = 56.5 \ V \ \angle 40.1°$ (polar)
    $V_{Z_1} = 43.2 + j36.4 \ V$ (rectangular)

    d. $V_{Z_2} = 22.6 \ V \ \angle{-27.9°}$ (polar)
    $V_{Z_2} = 20 - j10.6 \ V$ (rectangular)

    e. $V_{Z_3} = 45.2 \ V \ \angle{-34.9°}$ (polar)
    $V_{Z_3} = 37.1 - j25.9 \ V$ (rectangular)

    f. $V_T = V_{Z_1} + V_{Z_2} + V_{Z_3}$
    $100 \ V \ \angle 0° = (43.2 + j36.4 \ V) +$
       $(20 - j10.6 \ V) + (37.1 - j25.9 \ V)$
    $100 \ V \ \angle 0° = 100.3 - j0.1 \ V = 100.3 \ \angle 0°$

4. Equation: $V_3 = V_A - (V_1 + V_2)$
    $V_A = 100 + j0 \ V$ (rectangular)
    $V_1 = 22.2 + j33.3 \ V$ (rectangular)
    $V_2 = 15.8 + j12.3 \ V$ (rectangular)
    $V_3 = (100 + j0) - ((22.2 + j33.3) +$
       $15.8 + j12.3))$
    $V_3 = 62 - j45.6 \ V$ (rectangular)
    $V_3 = 77 \ V \ \angle{-36.3°}$ (polar)

5. Equation: $I_3 = I_1 + I_2$
    $I_1 = 1.41 - j1.41 \ A$
    $I_2 = 1.27 + j2.72 \ A$
    $I_3 = 2.68 + j1.31 \ A$ (rectangular)
    $I_3 = 2.98 \ A \ \angle 26°$

6. Parallel admittance/impedance of Z_2 and Z_L with V_2 shorted:

$$Y_2 = 125 \text{ mS } \angle 15° \text{ (polar)}$$
$$Y_2 = 121 + j32 \text{ mS (rectangular)}$$
$$Y_L = 33.3 \text{ mS } \angle -53° \text{ (polar)}$$
$$Y_L = 20 - j26.6 \text{ mS (rectangular)}$$
$$Y_{P_1} = 141 + j5.4 \text{ mS (rectangular)}$$
$$Y_{P_1} = 141 \text{ mS } \angle 2.2° \text{ (polar)}$$
$$Z_{P_1} = 7.09 \text{ } \Omega \text{ } \angle -2.2° \text{ (polar)}$$
$$Z_{P_1} = 7.08 - j0.27 \text{ } \Omega \text{ (rectangular)}$$

Add Z_1 and Z_P in series for total impedance as seen from V_1 (when V_2 is shorted).

$$Z_1 = 7.07 + j7.07 \text{ } \Omega \text{ (rectangular)}$$
$$Z_{T_1} = 14.2 + j6.8 \text{ } \Omega \text{ (rectangular)}$$
$$Z_{T_1} = 15.7 \text{ } \Omega \text{ } \angle 25.6° \text{ (polar)}$$

Current from V_1:

$$I_{T_1} = 3.18 \text{ A } \angle -25.6°$$

Voltage drops from V_1:

$$V_{Z_1} = 31.8 \text{ V } \angle 19.4° \text{ (polar)}$$
$$V_{Z_1} = 30 + j10.6 \text{ V (rectangular)}$$
$$V_{L_1} = 20 - j10.6 \text{ V (subtracted from } V_1)$$
$$V_{L_1} = 22.6 \text{ V } \angle -25.1°$$
$$V_{Z_2} = 20 - j10.6 \text{ V}$$

Current through the load impedance from V_1:

$$I_1 = 0.75 \text{ A } \angle -78.1° \text{ (polar)}$$
$$I_1 = 0.155 - j0.734 \text{ A (rectangular)}$$

Parallel admittance/impedance of Z_1 and Z_L with V_1 shorted:

$$Y_1 = 100 \text{ mS } \angle -45° \text{ (polar)}$$
$$Y_1 = 70.7 - j70.7 \text{ mS (rectangular)}$$
$$Y_L = 20 - j26.6 \text{ mS (rectangular)}$$
$$Y_{P_2} = 90.7 - j97.3 \text{ mS (rectangular)}$$
$$Y_{P_2} = 133 \text{ mS } \angle -47° \text{ (polar)}$$
$$Z_{P_2} = 7.52 \text{ } \Omega \text{ } \angle 47° \text{ (polar)}$$
$$Z_{P_2} = 5.13 + j5.5 \text{ } \Omega \text{ (rectangular)}$$

Add Z_2 and Z_{P_2} in series for total impedance as seen from V_2 (when V_1 is shorted).

$$Z_2 = 7.73 - j2.07 \text{ } \Omega \text{ (rectangular)}$$
$$Z_{T_2} = 12.8 + j3.43 \text{ } \Omega \text{ (rectangular)}$$
$$Z_{T_2} = 13.3 \text{ } \Omega \text{ } \angle 15° \text{ (polar)}$$

Current from V_2:

$$I_2 = 2.26 \text{ A } \angle 5°$$

Voltage drops from V_2:

$$V_2 = 28.2 + j10.3 \text{ V (rectangular)}$$
$$V_{Z_2} = 18.1 \text{ V } \angle -10° \text{ (polar)}$$
$$V_{Z_2} = 17.8 - j3.14 \text{ V (rectangular)}$$
$$V_{L_2} = 10.4 + j7.16 \text{ V (subtracted from } V_2)$$
$$V_{L_2} = 12.6 \text{ V } \angle 34.5°$$

Current through the load impedance from V_2:

$$I_2 = 0.42 \text{ A } \angle -18.5° \text{ (polar)}$$
$$I_2 = 0.398 - j0.133 \text{ A (rectangular)}$$

Combined current through load impedance:

$$I_L = I_1 + I_2$$
$$I_L = (0.155 - j0.734 \text{ A}) + (0.398 - j0.133 \text{ A})$$
$$I_L = 0.553 - j0.867 \text{ A (rectangular)}$$
$$I_L = 1.03 \text{ A } \angle -57.5° \text{ (polar)}$$

Combined voltage through load impedance:

$$V_L = 30.9 \text{ V } \angle -4.5°$$

7. Use admittance to combine Z_1 and Z_3:
$$Y_1 = 100 \text{ mS } \angle -45° \text{ (polar)}$$
$$Y_1 = 70.7 - j70.7 \text{ mS}$$
$$Y_3 = 100 \text{ mS } \angle 45° \text{ (polar)}$$
$$Y_3 = 70.7 + j70.7 \text{ mS}$$
$$Y_P = 141.4 \text{ mS } \angle 0°$$
$$Z_P = 7.07 \text{ } \Omega \text{ } \angle 0° \text{ (polar)}$$
$$Z_P = 7.07 + j0 \text{ } \Omega \text{ (rectangular)}$$

Find Z_{TH} by adding Z_P with Z_2 in series:
$$Z_2 = 35.4 - j35.4 \text{ } \Omega$$
$$Z_{TH} = Z_2 + Z_P$$
$$Z_{TH} = 42.5 - j35.4 \text{ } \Omega \text{ (rectangular)}$$
$$Z_{TH} = 55.3 \text{ } \Omega \text{ } \angle -39.8° \text{ (polar)}$$

Find V_{TH} using the voltage divider formula:

Formula: $V_{TH} = V_a \times \dfrac{Z_3}{Z_1 + Z_3}$

$Z_1 = 7.07 + j7.07 \ \Omega$ (rectangular)

$Z_3 = 7.07 - j7.07 \ \Omega$ (rectangular)

$Z_{1,3} = 14.14 + j0 \ \Omega$ (rectangular)

$Z_{1,3} = 14.1 \ \Omega \ \angle 0°$ (polar)

Substitution:

$V_{TH} = 25 \text{ V} \ \angle 0° \times \dfrac{10 \ \Omega \ \angle{-45°}}{14.1 \ \Omega \ \angle 0°}$

Answer: $V_{TH} = 17.7 \text{ V} \ \angle{-45°}$

8. The numerator is common to the three formulas:
    $Z_A Z_B + Z_B Z_C + Z_A Z_C$
    $Z_A Z_B = 30 \text{ k}\Omega^2 \ \angle 55°$ (polar)
    $Z_A Z_B = 17.2 + j24.6 \text{ k}\Omega^2$ (rectangular)
    $Z_B Z_C = 20 \text{ k}\Omega^2 \ \angle 40°$ (polar)
    $Z_B Z_C = 15.3 + j12.9 \text{ k}\Omega^2$ (rectangular)
    $Z_A Z_C = 60 \text{ k}\Omega^2 \ \angle 75°$ (polar)
    $Z_A Z_C = 15.5 + j58 \text{ k}\Omega^2$ (rectangular)
    Numerator: $(17.2 + j24.6 \ \Omega^2) +$
    $(15.3 + j12.9 \ \Omega^2) + (15.5 + j58 \ \Omega^2)$
    Numerator: $48 + j95.5 \text{ k}\Omega^2$ (rectangular)
    Numerator: $107 \text{ k}\Omega^2 \ \angle 63.3°$

Find Z_X:

Formula: $Z_X = \dfrac{\text{numerator}}{Z_A}$

Substitution: $Z_X = \dfrac{107 \text{ k}\Omega^2 \ \angle 63.3°}{300 \ \Omega \ \angle 45°}$

Answer: $Z_X = 357 \ \Omega \ \angle 18.3°$

Find Z_Y:

Formula: $Z_Y = \dfrac{\text{numerator}}{Z_B}$

Substitution: $Z_Y = \dfrac{107 \text{ k}\Omega^2 \ \angle 63.3°}{100 \ \Omega \ \angle 10°}$

Answer: $Z_Y = 1070 \ \Omega \ \angle 53.3°$

Find Z_Z:

Formula: $Z_Z = \dfrac{\text{numerator}}{Z_C}$

Substitution: $Z_Z = \dfrac{107 \text{ k}\Omega^2 \ \angle 63.3°}{200 \ \Omega \ \angle 30°}$

Answer: $Z_Z = 535 \ \Omega \ \angle 33.3°$

9. Denominator is common to the three formulas:
    $Z_X + Z_Y + Z_Z$
    $Z_X = 43.3 - j25 \ \Omega$
    $Z_Y = 10 + j17.3 \ \Omega$
    $Z_Z = 7.07 + j7.07 \ \Omega$
    Denominator: $60.4 - j0.63 \ \Omega$ (rectangular)
    Denominator: $60.4 \ \Omega \ \angle 0°$ (rounded for simplicity)

Find Z_A:

Formula: $Z_A = \dfrac{Z_Y Z_Z}{60.4 \ \Omega \ \angle 0°}$

Substitution:

$Z_A = \dfrac{(20 \ \Omega \ \angle 60°)(10 \ \Omega \ \angle 45°)}{60.4 \ \Omega \ \angle 0°}$

Answer: $Z_A = 3.31 \ \Omega \ \angle 105°$

Find Z_B:

Formula: $Z_B = \dfrac{Z_X Z_Z}{60.4 \ \Omega \ \angle 0°}$

Substitution:

$Z_B = \dfrac{(50 \ \Omega \ \angle{-30°})(10 \ \Omega \ \angle 45°)}{60.4 \ \Omega \ \angle 0°}$

Answer: $Z_B = 8.28 \ \Omega \ \angle 15°$

Find Z_C:

Formula: $Z_C = \dfrac{Z_X Z_Y}{60.4 \ \Omega \ \angle 0°}$

Substitution:

$Z_C = \dfrac{(50 \ \Omega \ \angle{-30°})(20 \ \Omega \ \angle 60°)}{60.4 \ \Omega \ \angle 0°}$

Answer: $Z_C = 16.6 \ \Omega \ \angle 30°$

ANSWERS TO STUDY GUIDE

Pages 189–202

1. a. $Z_1 = 141 + j141 \ \Omega$

    b. $Z_2 = 25 - j43.3 \ \Omega$

    c.   Total impedance:
         Formula: $Z_T = Z_1 + Z_2$
         Substitution:
         $\quad Z_T = (141 + j141 \ \Omega) + (25 - j43.3 \ \Omega)$
         Answer: $Z_T = 166 + j97.7 \ \Omega$

    d. $Z_T = 193 \ \Omega \ \angle 30.5°$

2. a. $Z_1 = 96.6 + j25.9 \ \Omega$

    b. $Z_2 = 12.9 - j48.3 \ \Omega$

    c. $Z_3 = 42.5 + j73.6 \ \Omega$

    d.   Total impedance:
         Formula: $Z_T = Z_1 + Z_2 + Z_3$
         Substitution: $Z_T = (96.6 + j25.9 \ \Omega) +$
         $\quad (12.9 - j48.3 \ \Omega) + (42.5 + j73.6 \ \Omega)$
         $Z_T = 152 + j51.2 \ \Omega$

    e. $Z_T = 160 \ \Omega \ \angle 18.6°$

3. Formula: $\dfrac{1}{Z_T} = \dfrac{1}{Z_1} + \dfrac{1}{Z_2}$

    Substitution: Use admittances.
    $\quad Y_1 = 20 \text{ mS} \ \angle 45° \text{ (polar)}$
    $\quad Y_1 = 14.1 + j14.1 \text{ mS (rectangular)}$
    $\quad Y_2 = 20 \text{ mS} \ \angle -60° \text{ (polar)}$
    $\quad Y_2 = 10 - j17.3 \text{ mS (rectangular)}$

    Admittance: $Y_T = 24.1 - j3.2 \text{ mS (rectangular)}$
    $\quad\quad\quad\quad\quad Y_T = 24.3 \text{ mS} \ \angle -7.5° \text{ (polar)}$

    Intermediate step: $Z_T = 1 \div 24.3 \text{ mS} \ \angle -7.5°$

    Answer: $Z_T = 41.2 \ \Omega \ \angle 7.5°$

4. Formula: $\dfrac{1}{Z_T} = \dfrac{1}{Z_1} + \dfrac{1}{Z_2}$

    Substitution: Use admittances.
    $\quad Y_1 = 100 \text{ mS} \ \angle 45° \text{ (polar)}$
    $\quad Y_1 = 70.7 + j70.7 \text{ mS (rectangular)}$
    $\quad Y_2 = 50 \text{ mS} \ \angle -30° \text{ (polar)}$
    $\quad Y_2 = 43.3 - j25 \text{ mS (rectangular)}$

    Admittance: $Y_T = 114 + j45.7 \text{ mS (rectangular)}$
    $\quad\quad\quad\quad\quad Y_T = 123 \text{ mS} \ \angle 21.8°$

    Intermediate step: $Z_T = 1 \div 123 \text{ mS} \ \angle 21.8°$

    Answer: $Z_T = 8.13 \ \Omega \ \angle -21.8°$

5. a.   Change to rectangular:
         $Z_1 = 30.1 + j39.9 \ \Omega$
         $Z_2 = 19.7 - j15.4 \ \Omega$
         $Z_3 = 70.7 - j70.7 \ \Omega$

    b.   Total impedance:
         Formula: $Z_T = Z_1 + Z_2 + Z_3$
         Substitution: $Z_T = (30.1 + j39.9 \ \Omega) +$
         $\quad (19.7 - j15.4 \ \Omega) + (70.7 - j70.7 \ \Omega)$
         Answer: $Z_T = 120.5 - j46.2 \ \Omega \text{ (rectangular)}$
         $\quad\quad\quad\quad Z_T = 129 \ \Omega \ \angle -21° \text{ (polar)}$

    c.   Circuit current:
         Formula: $I = V/Z$
         Substitution: $I = 100 \text{ V} \ \angle 0° \div 129 \ \Omega \ \angle 21°$
         Answer: $I = 0.775 \text{ A} \ \angle 21°$

    d.   Voltage across Z_1:
         Formula: $V_{Z_1} = I \times Z_1$
         Substitution:
         $\quad V_{Z_1} = 0.775 \text{ A} \ \angle 21° \times 50 \ \Omega \ \angle 53°$
         Answer: $V_{Z_1} = 38.8 \text{ V} \ \angle 74° \text{ (polar)}$

    e.   Voltage across Z_2:
         Formula: $V_{Z_2} = I \times Z_2$
         Substitution:
         $\quad V_{Z_2} = 0.775 \text{ A} \ \angle 21° \times 25 \ \Omega \ \angle -38°$
         Answer: $V_{Z_2} = 19.4 \text{ V} \ \angle -17° \text{ (polar)}$

f. Voltage across Z_3:

Formula: $V_{Z_3} = I \times Z_3$

Substitution:

$V_{Z_3} = 0.775$ A $\angle 21° \times 100$ Ω $\angle{-45}°$

Answer: $V_{Z_1} = 77.5$ V $\angle{-23}°$ (polar)

g. $V_{Z_1} = 10.7 + j37.3$ V (rectangular)

$V_{Z_2} = 18.6 - j5.7$ V (rectangular)

$V_{Z_3} = 71.3 - j30.3$ V (rectangular)

h. Student labeled figure.

i. Kirchhoff's voltage law:

Formula: $V_T = V_{Z_1} + V_{Z_2} + V_{Z_3}$

Substitution: 100 V $\angle 0° \cong (10.7 + j37.3) +$
(18.6 − j5.7) + (71.3 − j30.3)

Answer: 100 V $\angle 0° \cong 100.6 + j1.3$ V

Sum of the voltages in polar: 100.6 V $\angle 0.74°$

6. a. Change to rectangular:

$V_a = 100 + j0$ V

$V_1 = 42.4 - j42.4$ V

$V_3 = 35.4 + j35.4$ V

b. Solve using Kirchhoff's law:

Formula: $V_2 = V_a - (V_1 + V_3)$

Substitution: $V_2 = (100 + j0$ V$) -$
[(42.4 − j42.4 V) + (35.4 + j35.4 V)]

$V_2 = 22.2 + j7$ V (rectangular)

Answer: $V_2 = 23.3$ V $\angle 17.5°$ (polar)

7. Formula: $I_3 = I_1 + I_2$

I_1 rectangular: $I_1 = 0.94 + 0.342$ A

I_2 rectangular: $I_2 = 1.41 + j1.41$ A

Substitution: $I_3 = 2.35 + j1.75$ A (rectangular)

Answer: $I_3 = 2.93$ A $\angle 36.7°$

8. *Part 1—V_2 shorted:*

a. Student sketch of equivalent circuit.

b. Parallel admittance/impedance of Z_2 and Z_L
with V_2 shorted:

Formula: $Z_{P_1} = \dfrac{1}{Y_{P_1}}$ and $Y_{P_1} = Y_2 + Y_L$

$Y_2 = 125$ mS $\angle 45°$ (polar)

$Y_2 = 88.4 + j88.4$ mS (rectangular)

$Y_L = 100$ mS $\angle 15°$ (polar)

$Y_L = 96.6 + j25.9$ mS (rectangular)

Substitution:

$Y_{P_1} = (88.4 + j88.4$ mS$) + (96.6 + j25.9$ mS$)$

$Y_{P_1} = 185 + j114$ mS (rectangular)

$Y_{P_1} = 217$ mS $\angle 31.6°$ (polar)

Answer: $Z_{P_1} = 4.61$ Ω $\angle{-31.6}°$ (polar)

$Z_{P_1} = 3.93 - j2.42$ Ω (rectangular)

c. Add Z_1 and Z_{P_1} in series for total impedance
as seen from V_1 (when V_2 is shorted).

Formula: $Z_{T_1} = Z_1 + Z_{P_1}$

$Z_1 = 2 + j3.46$ Ω (rectangular)

Substitution:

$Z_{T_1} = (2 + j3.46$ Ω$) + (3.93 - j2.42$ Ω$)$

Answer: $Z_{T_1} = 5.93 + j1.04$ Ω (rectangular)

$Z_{T_1} = 6.02$ Ω $\angle 10°$ (polar)

d. Current from V_1:

Formula: $I_{T_1} = V_1 \div Z_{T_1}$

Substitution: $I_{T_1} = 10$ V $\angle 0° \div 6.02$ Ω $/10°$

Answer: $I_{T_1} = 1.66$ A $\angle{-10}°$

e. Voltage drop across Z_1, from V_1:

Formula: $V_{Z_1} = I_{T_1} \times Z_1$

Substitution: $V_{Z_1} = 1.66$ A $\angle{-10}° \times 4$ Ω $\angle 60°$

Answer: $V_{Z_1} = 6.64$ V $\angle 50°$ (polar)

$V_{Z_1} = 4.27 + j5.09$ V (rectangular)

f. Voltage drop across Z_L, from V_1:

Formula: $V_{L_1} = I_{T_1} \times Z_{P_1}$

Substitution:

$V_{L_1} = 1.66$ A $\angle{-10}° \times 4.61$ Ω $\angle{-31.6}°$

Answer: $V_{L_1} = 7.65$ V $\angle{-41.6}°$

g. Current through the load impedance from V_1:

Formula: $I_1 = V_{L_1} \div Z_L$

Substitution:

$I_1 = 7.65$ V $\angle{-41.6}° \div 10$ Ω $\angle{-15}°$

Answer: $I_1 = 0.765$ A $\angle{-26.6}°$ (polar)

$I_1 = 0.684 - j0.343$ A (rectangular)

Part 2—V_1 shorted:

a.   Student sketch of equivalent circuit.

b.   Parallel admittance/impedance of Z_1 and Z_L
     with V_1 shorted:

   Formula: $Z_{P_2} = \dfrac{1}{Y_{P_2}}$ and $Y_{P_2} = Y_1 + Y_L$

   $Y_1 = 250$ mS $\angle{-60°}$ (polar)

   $Y_1 = 125 - j217$ mS (rectangular)

   $Y_L = 96.6 + j25.9$ mS (rectangular)

   Substitution:

   $Y_{P_2} = (125 - j217$ mS$) + (96.6 + j25.9$ mS$)$

   $Y_{P_2} = 222 - j191$ mS (rectangular)

   $Y_{P_2} = 293$ mS $\angle{-40.7°}$ (polar)

   Answer: $Z_{P_2} = 3.41$ Ω $\angle{40.7°}$ (polar)

   $Z_{P_2} = 2.59 + j2.22$ Ω (rectangular)

c.   Add Z_2 and Z_{P_2} in series for total impedance
     as seen from V_2 (when V_1 is shorted).

   Formula: $Z_{T_2} = Z_2 + Z_{P_1}$

   $Z_2 = 5.66 - j5.66$ Ω (rectangular)

   Substitution:

   $Z_{T_2} = (5.66 + j5.66$ Ω$) + (2.59 - j2.22$ Ω$)$

   Answer: $Z_{T_2} = 8.25 - j3.44$ Ω (rectangular)

   $Z_{T_2} = 8.94$ Ω $\angle{-22.6°}$ (polar)

d.   Current from V_2:

   Formula: $I_{T_2} = V_2 \div Z_{T_2}$

   Substitution:

   $I_{T_2} = 15$ V $\angle{-15°} \div 8.94$ Ω $\angle{-22.6°}$

   Answer: $I_{T_2} = 1.68$ A $\angle{7.6°}$

e.   Voltage drop across Z_2, from V_2:

   Formula: $V_{Z_2} = I_{T_2} \times Z_2$

   Substitution:

   $V_{Z_2} = 1.68$ A $\angle{7.6°} \times 8$ Ω $\angle{-45°}$

   Answer: $V_{Z_2} = 13.4$ V $\angle{-37.4°}$ (polar)

   $V_{Z_2} = 10.6 - j8.14$ V (rectangular)

f.   Voltage drop across Z_L, from V_2:

   Formula: $V_{L_2} = I_{T_2} \times Z_{P_2}$

   Substitution:

   $V_{L_2} = 1.68$ A $\angle{7.6°} \times 3.41$ Ω $\angle{-40.7°}$

   Answer: $V_{L_2} = 5.73$ V $\angle{48.3°}$

g.   Current through the load impedance from V_2:

   Formula: $I_2 = V_{L_2} \div Z_L$

   Substitution:

   $I_1 = 5.73$ V $\angle{48.3°} \div 10$ Ω $\angle{-15°}$

   Answer: $I_2 = 0.573$ A $\angle{63.3°}$ (polar)

   $I_2 = 0.257 + j0.512$ A (rectangular)

Part 3—Combined results.

a.   Combined current through load impedance:
     Formula: $I_L = I_1 + I_2$
     Substitution: $I_L = (0.684 - j0.343$ A$) +$
     $(0.257 + j0.512$ A$)$
     Answer: $I_L = 0.941 + j0.169$ A (rectangular)
     $I_L = 0.956$ A $\angle{10.2°}$ (polar)

b.   Combined voltage through load impedance:
     Formula: $V_L = I_L \times Z_L$
     Substitution:
     $V_L = 0.956$ A $\angle{10.2°} \times 10$ Ω $\angle{-15°}$
     Answer: $V_L = 9.56$ V $\angle{-4.8°}$

9. a.   Parallel combination of Z_1 and Z_3:
         Formula: Use admittance to combine Z_1
         and Z_3:

   Formula: $Z_P = \dfrac{1}{Y_P}$ and $Y_P = Y_1 + Y_3$

   $Y_1 = 50$ mS $\angle{-45°}$ (polar)
   $Y_1 = 35.4 - j35.4$ mS (rectangular)
   $Y_3 = 35.4 - j35.4$ mS (polar)
   $Y_3 = 8.84 + j8.84$ mS

   Substitution:
   $Y_P = (35.4 - j35.4$ mS$) - (8.84 + j8.84$ mS$)$
   $Y_P = 44.2 - j26.6$ mS (rectangular)
   $Y_P = 51.6$ mS $\angle{31°}$ (polar)

   Answer: $Z_P = 19.4$ Ω $\angle{-31°}$ (polar)
   $Z_P = 16.6 - j10$ Ω (rectangular)

b.   Find Z_{TH} by adding Z_P with Z_2 in series.

$Z_2 = 6 + j8 \ \Omega$

$Z_P = 16.6 - j10 \ \Omega$

Formula: $Z_{TH} = Z_2 + Z_P$

Substitution: $Z_{TH} = (6 + j8 \ \Omega) + (16.6 - j10 \ \Omega)$

Answer: $Z_{TH} = 22.6 - j2 \ \Omega$ (rectangular)

$Z_{TH} = 22.7 \ \Omega \ \angle{-5.1°}$ (polar)

c.   Find V_{TH} using the voltage divider formula:

Formula: $V_{TH} = V_a \times \dfrac{Z_3}{Z_1 + Z_3}$

$Z_1 = 14.1 + j14.1 \ \Omega$ (rectangular)

$Z_3 = 56.6 - j56.6 \ \Omega$ (rectangular)

$Z_1 + Z_3 = 70.7 - j42.5 \ \Omega$ (rectangular)

Denominator $= 82.5 \ \Omega \ \angle{-31°}$ (polar)

Substitution:

$V_{TH} = 50 \ V \ \angle{0°} \times \dfrac{80 \ \Omega \ \angle{-45°}}{82.5 \ \Omega \ \angle{-31°}}$

Answer: $V_{TH} = 48.5 \ V \ \angle{-14°}$

10. a.   Solve the numerator:

Formula: Numerator $= Z_A Z_B + Z_B Z_C + Z_A Z_C$

$Z_A Z_B = 5000 \ \Omega^2 \ \angle{-20°}$ (polar)

$Z_A Z_B = 4700 - j1710 \ \Omega^2$ (rectangular)

$Z_B Z_C = 5000 \ \Omega^2 \ \angle{70°}$ (polar)

$Z_B Z_C = 1710 + j4700 \ \Omega^2$ (rectangular)

$Z_A Z_C = 2500 \ \Omega^2 \ \angle{30°}$ (polar)

$Z_A Z_C = 2170 + j1250 \ \Omega^2$ (rectangular)

Substitution: Numerator $= (4700 - j1710 \ \Omega^2) +$

$(1710 + j4700 \ \Omega^2) + (2170 + j1250 \ \Omega^2)$

Answer: Numerator: $8580 + j4240 \ \Omega^2$

(rectangular)

Numerator: $9570 \ \Omega^2 \ \angle{26.3°}$

b.   Find Z_X:

Formula: $Z_X = \dfrac{\text{numerator}}{Z_A}$

Substitution: $Z_X = \dfrac{9570 \ \Omega^2 \ \angle{26.3°}}{50 \ \Omega \ \angle{-30°}}$

Answer: $Z_X = 191 \ \Omega \ \angle{56.3°}$

c.   Find Z_Y:

Formula: $Z_Y = \dfrac{\text{numerator}}{Z_B}$

Substitution: $Z_Y = \dfrac{9570 \ \Omega^2 \ \angle{26.3°}}{100 \ \Omega \ \angle{10°}}$

Answer: $Z_Y = 95.7 \ \Omega \ \angle{16.3°}$

d.   Find Z_Z:

Formula: $Z_Z = \dfrac{\text{numerator}}{Z_C}$

Substitution: $Z_Z = \dfrac{9570 \ \Omega^2 \ \angle{26.3°}}{50 \ \Omega \ \angle{60°}}$

Answer: $Z_Z = 191 \ \Omega \ \angle{-33.7°}$

11. a.   Solve the denominator:

Formula: Denominator $= Z_X + Z_Y + Z_Z$

$Z_X = 8.66 + j5 \ \Omega$

$Z_Y = 14.1 - j14.1 \ \Omega$

$Z_Z = 15 + j26 \ \Omega$

Substitution: Denominator $= (8.66 + j5 \ \Omega) +$

$(14.1 - j14.1 \ \Omega) + (15 + j26 \ \Omega)$

Answer: Denominator $= 37.8 + j16.9 \ \Omega$

(rectangular)

Denominator: $41.4 \ \Omega \ \angle{24°}$ (polar)

b.   Find Z_A:

Formula: $Z_A = \dfrac{Z_Y Z_Z}{\text{denominator}}$

Substitution:

$Z_A = \dfrac{(20 \ \Omega \ \angle{-45°})(30 \ \Omega \ \angle{60°})}{41.4 \ \Omega \ \angle{24°}}$

Answer: $Z_A = 14.5 \ \Omega \ \angle{-9°}$

c.   Find Z_B:

Formula: $Z_B = \dfrac{Z_X Z_Z}{41.4 \ \Omega \ \angle{24°}}$

Substitution:

$Z_B = \dfrac{(10 \ \Omega \ \angle{30°})(30 \ \Omega \ \angle{60°})}{41.4 \ \Omega \ \angle{24°}}$

Answer: $Z_B = 7.25 \ \Omega \ \angle{66°}$

d. Find Z_C:

Formula: $Z_C = \dfrac{Z_X Z_Y}{\text{denominator}}$

Substitution:

$$Z_C = \frac{(10\ \Omega\ \angle 30°)\ (20\ \Omega\ \angle{-45°})}{41.4\ \Omega\ \angle 24°}$$

Answer: $Z_C = 4.83\ \Omega\ \angle{-39°}$

ANSWERS TO CHAPTER TEST IN THE INSTRUCTOR'S MANUAL

Pages 269–282

1. The sum of voltages around a closed loop equals zero.

2. $R = Z \times \cos\theta$
    $X = Z \times \sin\theta$

    a. $Z_1 = 106 + j106\ \Omega$

    b. $Z_1 = 125 - j217\ \Omega$

    c. Formula: $Z_T = Z_1 + Z_2$
        Substitution:
        $Z_T = (106 + j106\ \Omega) + (125 - j217\ \Omega)$
        Answer: $Z_T = 231 - j111\ \Omega$ (rectangular)

    d. $Z_T = 256\ \Omega\ \angle{-25.7°}$ (polar)

3. Formula: $\dfrac{1}{Z_T} = \dfrac{1}{Z_1} + \dfrac{1}{Z_2}$

    Substitution: $\dfrac{1}{Z_T} = \dfrac{1}{10\ \Omega\ \angle{-45°}} + \dfrac{1}{10\ \Omega\ \angle 60°}$

    Admittance: $\dfrac{1}{Z_T} = 0.1\ \text{S}\ \angle 45° + 0.1\ \text{S}\ \angle{-60°}$

    Rectangular: $\dfrac{1}{Z_T} = (0.0707 + j0.0707\ \text{S}) +$
    $(0.05 - j0.0866\ \text{S})$

    Intermediate step: $\dfrac{1}{Z_T} = 0.1207 - j0.0159\ \text{S}$
    $\dfrac{1}{Z_T} = 0.122\ \text{S}\ \angle{-7.5°}$

    Answer: $Z_T = 8.2\ \Omega\ \angle 7.5°$

4. a. Change all given voltages to rectangular form.
        $V_a = 50 + j0$ V
        $V_1 = 7.07 - j7.07$ V
        $V_3 = 14.14 + j14.14$ V

    b. Use Kirchhoff's voltage law.
        Kirchhoff's equation: $V_2 = V_a - (V_1 + V_2)$
        Substitution: $V_2 = (50 + j0$ V$) - [(7.07 - j7.07$ V$) + (14.14 + j14.14$ V$)]$
        Intermediate step: $V_2 = (50 + j0$ V$) - (21.21 + j7.07$ V$)$
        Answer (rectangular): $V_2 = 28.79 - j7.07$ V
        Answer (polar): $V_2 = 29.6$ V $\angle{-13.8°}$ (polar)

5. I_1 (rectangular) = $4.33 + j2.5$ A
    I_2 (rectangular) = $2.12 + j2.12$ A
    Equation: $I_3 = I_1 + I_2$
    Substitution: $I_3 = (4.33 + j2.5$ A$) + (2.12 + j2.12$ A$)$
    Answer (rectangular): $I_3 = 6.45 + j4.62$ A
    Answer (polar): $I_3 = 7.93$ A $\angle 35.6°$

6. a.

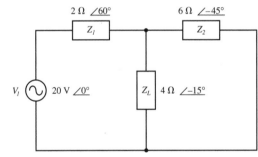

    b. Find the parallel combination of Z_2 and Z_L.

        Formula: $\dfrac{1}{Z_{P_1}} = \dfrac{1}{Z_2} + \dfrac{1}{Z_L}$

        Substitution:

        $$\frac{1}{Z_{P_1}} = \frac{1}{6\ \Omega\ \angle{-45°}} + \frac{1}{4\ \Omega\ \angle{-15°}}$$

        Admittance:

        $$\frac{1}{Z_{P_1}} = 0.167\ \text{S}\ \angle 45° + 0.25\ \text{S}\ \angle 15°$$

        Admittance (rectangular):

        $$\frac{1}{Z_{P_1}} = (0.118 + j0.118\ \text{S}) + (0.241 + j0.0647\ \text{S})$$

Combining terms :

$$\frac{1}{Z_{P_1}} = 0.359 + j0.183 \text{ S} = 0.403 \text{ S } \angle 27°$$

Answer (polar): $Z_{P_1} = 2.48 \text{ }\Omega \angle -27°$

Answer (rectangular): $Z_{P_1} = 2.21 - j1.13 \text{ }\Omega$

c. Combine the parallel equivalent with the series Z_1.

Convert Z_1 rectangular: $Z_1 = 1 + j1.73 \text{ }\Omega$

Formula: $Z_{T_1} = Z_1 + Z_{P_1}$

Substitution: $Z_{T_1} = (1 + j1.73 \text{ }\Omega) + (2.21 - j1.13 \text{ }\Omega)$

Answer (rectangular): $Z_{T_1} = 3.21 + j0.6 \text{ }\Omega$

Answer (polar): $Z_{T_1} = 3.27 \text{ }\Omega \angle 10.6°$

7. a. Total current supplied by V_1:

Formula: $I_{T_1} = \dfrac{V_1}{Z_{T_1}}$

Substitution: $I_{T_1} = \dfrac{20 \text{ V } \angle 0°}{3.27 \text{ }\Omega \angle 10.6°}$

Answer: $I_{T_1} = 6.12 \text{ A } \angle -10.6°$

b. Voltage drop across the series impedance Z_1:

Formula: $V_{Z_1} = I_{T_1} \times Z_1$

Substitution: $V_{Z_1} = 6.12 \text{ A } \angle -10.6° \times 2 \text{ }\Omega \angle 60°$

Answer: $V_{Z_1} = 12.24 \text{ V } \angle 49.4°$

c. Voltage drop across the load, supplied by V_1:

Formula: $V_{L_1} = I_{T_1} \times Z_{P_1}$

Substitution: $V_{L_1} = 6.12 \text{ A } \angle -10.6° \times 2.48 \text{ }\Omega \angle -27°$

Answer: $V_{L_1} = 15.18 \text{ V } \angle -37.6°$

d. Current through the load supplied by V_1:

Formula: $I_{L_1} = \dfrac{V_{L_1}}{Z_L}$

Substitution: $I_{L_1} = \dfrac{15.18 \text{ V } \angle -37.6°}{4 \text{ }\Omega \angle -15°}$

Answer (polar): $I_{L_1} = 3.8 \text{ A } \angle -22.6°$

Answer (rectangular): $I_{L_1} = 3.51 - j1.46 \text{ A}$

8. a. Draw the equivalent circuit here for reference.

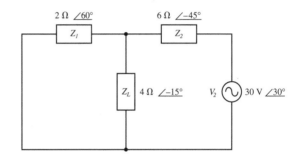

b. Find the parallel combination of Z_1 and Z_L.

Formula: $\dfrac{1}{Z_{P_2}} = \dfrac{1}{Z_1} + \dfrac{1}{Z_L}$

Substitution:

$$\frac{1}{Z_{P_2}} = \frac{1}{2 \text{ }\Omega \angle 60°} + \frac{1}{4 \text{ }\Omega \angle -15°}$$

Admittance:

$$\frac{1}{Z_{P_2}} = 0.5 \text{ S } \angle -60° + 0.25 \text{ S } \angle 15°$$

Rectangular:

$$\frac{1}{Z_{P_2}} = (0.25 - j0.433 \text{ S}) + (0.241 + j0.0647 \text{ S})$$

Combining terms:

$$\frac{1}{Z_{P_2}} = 0.491 - j0.368 \text{ S} = 0.614 \text{ S } \angle -36.9°$$

Answer (polar): $Z_{P_2} = 1.63 \text{ }\Omega \angle 36.9°$

Answer (rectangular): $Z_{P_2} = 1.3 + j0.979 \text{ }\Omega$

c. Combine the parallel equivalent with the series Z_2.

Convert Z_2 to rectangular: $Z_2 = 4.24 - j4.24\ \Omega$

Formula: $Z_{T_2} = Z_2 + Z_{P_2}$

Substitution: $Z_{T_2} = (4.24 - j4.24\ \Omega) + (1.3 + j0.979\ \Omega)$

Answer (rectangular): $Z_{T_2} = 5.54 - j3.26\ \Omega$

Answer (polar): $Z_{T_2} = 6.43\ \Omega\ \angle{-30.5°}$

9. a. Total current supplied by V_2:

Formula: $I_{T_2} = \dfrac{V_2}{Z_{T_2}}$

Substitution: $I_{T_2} = \dfrac{30\ V\ \angle 30°}{6.43\ \Omega\ \angle{-30.5°}}$

Answer: $I_{T_2} = 4.67\ A\ \angle 60.5°$

b. Voltage drop across the series impedance Z_2:

Formula: $V_{Z_2} = I_{T_2} \times Z_2$

Substitution: $V_{Z_2} = 4.67\ A\ \angle 60.5° \times 6\ \Omega\ \angle{-45°}$

Answer: $V_{Z_2} = 28.0\ V\ \angle 15.5°$

c. Voltage drop across the load, supplied by V_2:

Formula: $V_{L_2} = I_{T_2} \times Z_{P_2}$

Substitution: $V_{L_2} = 4.67\ A\ \angle 60.5° \times 1.63\ \Omega\ \angle 36.9°$

Answer: $V_{L_2} = 7.61\ \angle 97.4°$

d. Current through the load supplied by V_2:

Formula: $I_{L_2} = \dfrac{V_{L_2}}{Z_L}$

Substitution: $I_{L_2} = \dfrac{7.61\ V\ \angle 97.4°}{4\ \Omega\ \angle{-15°}}$

Answer (polar): $I_{L_2} = 1.90\ A\ \angle 112.4°$

Answer (rectangular): $I_{L_2} = -0.724 - j1.76\ A$

10. a. Combined current through the load:

Formula: $I_{L_T} = I_{L_1} + I_{L_2}$

Substitution: $I_{L_T} = (3.51 - j1.46\ A) + (-0.724 - j1.76\ A)$

Answer (rectangular): $I_{L_T} = 2.79 - j3.22\ A$

Answer (polar): $I_{L_T} = 4.26\ A\ \angle{-49.1°}$

b. Ohm's law to find the voltage across the load.

Formula: $V_{L_T} = I_{L_T} \times Z_L$

Substitution: $V_{L_T} = 4.26\ A\ \angle{-49.1°} \times 4\ \Omega\ \angle{-15°}$

Answer (polar): $V_{L_T} = 17.0\ V\ \angle 64.1°$

Answer (rectangular): $V_{L_T} = 7.43 + j15.3\ V$

Chapter Test

21

Circuit Theorems Applied to AC

1. Kirchhoff's voltage law states: _____

   _____

2. Find the total impedance of two networks connected in series with values $Z_1 = 150 \ \Omega \ \angle 45°$ and $Z_2 = 250 \ \Omega \ \angle -60°$.

   a. Convert Z_1 to rectangular form.

      Z_1 (rectangular) = _____

   b. Convert Z_2 to rectangular form.

      Z_2 (rectangular) = _____

   c. Total impedance:

      Formula: _____

      Substitution: _____

      Answer (rectangular): _____

   d. Change Z_T to polar form.

      Z_T (polar) = _____

3. Find the total impedance of two networks connected in parallel with impedance values $Z_1 = 10 \ \Omega \ \angle{-45°}$ and $Z_2 = 10 \ \Omega \ \angle{60°}$.

Formula: _____

Substitution: _____

Admittance: _____

Rectangular form: _____

Intermediate step: _____

Answer: _____

4. Find the missing voltage drop in the figure shown below.

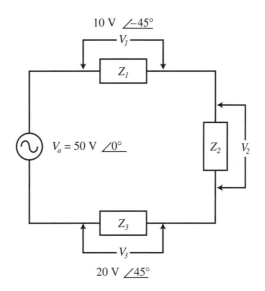

a. Change all given voltages to rectangular form.

V_a = _____

V_1 = _____

V_3 = _____

b. Use Kirchhoff's voltage law.

Kirchhoff's equation: _____

Intermediate step: _____

Substitution: _____

Answer (rectangular): V_2 = _____

Answer (polar): V_2 = _____

5. Use Kirchhoff's current law to find current I_3 in the figure.

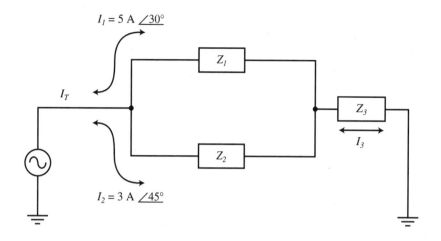

I_1 (rectangular) = _____

I_2 (rectangular) = _____

Kirchhoff's equation: _____

Substitution: _____

Answer (rectangular): I_3 = _____

Answer (polar): I_3 = _____

For questions 6 through 10, use the superposition theorem to solve the ac circuit shown in the following figure.

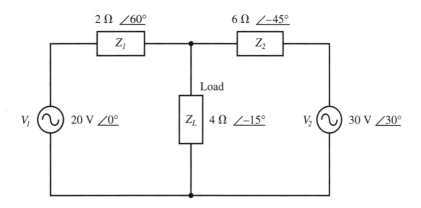

6. Find the total impedance as seen from V_1 with V_2 shorted.

   a. Draw the equivalent circuit here for reference.

b. Find the parallel combination of Z_2 and Z_L.

Formula: _____

Substitution: _____

Admittance: _____

Admittance (rectangular): _____

Combining terms: _____

Answer (polar): _____

Answer (rectangular): _____

c. Combine the parallel equivalent with the series Z_1.

Convert Z_1 to rectangular: _____

Formula: _____

Substitution: _____

Answer (rectangular): _____

Answer (polar): _____

7. Using the circuit drawn in part a of question 6, find the values requested.

    a. Total current supplied by V_1:

        Formula: _____

        Substitution: _____

        Answer: _____

    b. Voltage drop across the series impedance Z_1:

        Formula: _____

        Substitution: _____

        Answer: _____

c. Voltage drop across the load, supplied by V_l:

Formula: _____

Substitution: _____

Answer: _____

d. Current through the load supplied by V_l:

Formula: _____

Substitution: _____

Answer (polar): _____

Answer (rectangular): _____

8. Return to the original circuit of question 6. Solve for total impedance as seen from V_2 with V_1 shorted.

    a. Draw the equivalent circuit here for reference.

    b. Find the parallel combination of Z_1 and Z_L.

        Formula: _____

        Substitution: _____

        Admittance: _____

        Admittance (rectangular): _____

        Combining terms: _____

        Answer (polar): _____

        Answer (rectangular): _____

c. Combine the parallel equivalent with the series Z_2.

Convert Z_2 to rectangular: _____

Formula: _____

Substitution: _____

Answer (rectangular): _____

Answer (polar): _____

9. Using the circuit drawn in part a of question 8, find the values requested.

a. Total current supplied by V_2:

Formula: _____

Substitution: _____

Answer: _____

b. Voltage drop across the series impedance Z_2:

Formula: _____

Substitution: _____

Answer: _____

c. Voltage drop across the load, supplied by V_2:

Formula: _____

Substitution: _____

Answer: _____

d. Current through the load supplied by V_2:

Formula: _____

Substitution: _____

Answer (polar): _____

Answer (rectangular): _____

10. Combined results at the load from questions 6 through 9.

a. Combined current through the load:

Formula: _____

Substitution: _____

Answer (rectangular): _____

Answer (polar): _____

b. Ohm's law to find the voltage across the load.

Formula: _____

Substitution: _____

Answer (polar): _____

Answer (rectangular): _____

Chapter 22

AC Power

OBJECTIVES

After studying this chapter, students should be able to:
- Describe the power curve in an ac circuit.
- State the difference between absorbed and reflected ac power.
- Describe ac power in a purely resistive circuit.
- Describe ac power in a purely inductive circuit.
- Describe ac power in a purely capacitive circuit.
- Describe ac power in a circuit with resistance and reactance.
- Draw and label the phasor triangle for ac power.
- Define real power, reactive power, and apparent power.
- Define power factor.
- In an ac circuit, calculate the values for: impedance, phase angle, current, individual voltage drops, and power.
- Describe the need for power factor correction.
- Calculate the value of a capacitor needed for power factor correction in an inductive circuit.

INSTRUCTIONAL MATERIALS

Text: Pages 667–682
　　　Test Your Knowledge Questions, Pages 681–682
Study Guide: Pages 203–208

ANSWERS TO TEXTBOOK

Test Your Knowledge, Pages 681–682

1. With pure resistance, the curves have voltage, current, and power all in phase. The full amount of power reaches the load.
2. With pure reactance, the voltage and current curves are 90° out of phase, resulting in a power curve that has as much power reflected from the load as absorbed. The net power to the load is zero.
3. Resistive.
4. Reactive.
5. real power, watts
6. reactive power, VAR
7. reactive power, VAR
8. apparent power, VA
9. Refer to figure 22-5.
10. Refer to figure 22-5.
11. Power factor is the ratio of real power to reactive power. There are no units. Its range is from 0 to 1.
12. For maximum efficiency, the power factor should be 1, which is purely resistive.
13. a. $Z = 90.1\ \Omega$
    b. $\theta = 56.3°$
    c. $I = 1.11$ A
    d. $V_R = 55.5$ V
    e. $V_L = 83.3$ V
    f. $P_R = 61.6$ W
    g. $P_Q = 92.5$ VAR
    h. $P_S = 111$ VA
    i. PF = 0.555
14. Capacitive.
15. Inductive.
16. a. Operating angle from power factor: $\theta = 41.4°$
    b. $P_Q = 882$ VAR
    c. $P_S = 1333$ VA
    d. $I = 11.1$ A
17. a. $I_C = 7.35$ A
    b. $X_C = 16.3\ \Omega$
    c. $C = 163\ \mu F$

ANSWERS TO STUDY GUIDE

Pages 203–208

1. a.
2. c.
3. a.
4. d.
5. a.

283

6. b.
7. b.
8. c.
9. d.
10. Refer to textbook figure 22-5.
11. Refer to textbook figure 22-5.
12. b.

13. a. Impedance:

Formula: $Z = \sqrt{R^2 + X_L^2}$

Substitution: $Z = \sqrt{100^2 + 150^2}$

Answer: $Z = 180\ \Omega$

b. Phase angle:

Formula: $\theta = \tan^{-1} \dfrac{X_L}{R}$

Substitution: $\theta = \tan^{-1} \dfrac{150\ \Omega}{100\ \Omega}$

Answer: $\theta = 56.3°$

c. Current:

Formula: $I = \dfrac{V_a}{Z}$

Substitution: $I = \dfrac{100\ \text{V}\ \angle 0°}{180\ \Omega\ \angle 56.3°}$

Answer: $I = 0.556\ \text{A}\ \angle{-56.3°}$

d. Voltage drops:
Formula: $V_R = I \times R$
Substitution:

$V_R = 0.556\ \text{A}\ \angle{-56.3°} \times 100\ \Omega\ \angle 0°$
Answer: $V_R = 55.6\ \text{V}\ \angle{-56.3°}$
$V_L = 83.4\ \text{V}\ \angle 33.7°$

e. Real power:
Formula: $P_R = I \times V_R$
Substitution: $P_R = 0.556\ \text{A} \times 55.6\ \text{V}$
Answer: $P_R = 30.9\ \text{W}$

f. Reactive power:
Formula: $P_Q = I \times V_L$
Substitution: $P_Q = 0.556\ \text{A} \times 83.4\ \text{V}$
Answer: $P_Q = 46.4\ \text{VAR}$

g. Apparent power:
Formula: $P_S = I \times V$
Substitution: $P_S = 0.556\ \text{A} \times 100\ \text{V}$
Answer: $P_S = 55.6\ \text{VA}$

h. Power factor:
Formula: $\text{PF} = \cos\theta$
Substitution: $\text{PF} = \cos 56.3°$
Answer: $\text{PF} = 0.555 = 55.5\%$

14. c.
15. b.
16. a. Student sketch of equivalent circuit.
    b. Phase angle:
Formula: $\theta = \cos^{-1} \text{PF}$
Substitution: $\theta = \cos^{-1} 0.85$
Answer: $\theta = 31.8°$
    c. Reactive power:
Formula: $P_Q = P_R \times \tan\theta$
Substitution: $P_Q = 2000\ \text{W} \times \tan 31.8°$
Answer: $P_Q = 1240\ \text{VAR}$
    d. Apparent power:
Formula: $P_S = \sqrt{P_R^2 + P_Q^2}$
Substitution: $P_S = \sqrt{2000^2 + 1240^2}$
Answer: $P_S = 2353\ \text{VA}$
    e. Student sketch of power triangle.

    f. Current:

Formula: $I = \dfrac{P_S}{V_a}$

Substitution: $I = \dfrac{2353\ \text{VA}}{120\ \text{V}}$

Answer: $I = 19.6\ \text{A}$

17. a. Reactive power in the capacitor:
Capacitive power equals inductive power:
$P_Q = 1240\ \text{VAR}$

    b. Capacitive current:

Formula: $I_C = \dfrac{P_Q}{V_a}$

Substitution: $I_C = \dfrac{1240\ \text{VAR}}{120\ \text{V}}$

Answer: $I_C = 10.3\ \text{A}$

c.   Capacitive reactance:

Formula: $X_C = \dfrac{V_a}{I_C}$

Substitution: $X_C = \dfrac{120 \text{ V}}{10.3 \text{ A}}$

Answer: $X_C = 11.7 \text{ }\Omega$

d.   Circuit current after PF correction:

Formula: $I = \dfrac{P_S}{V_a}$

Substitution: $I = \dfrac{2000 \text{ VA}}{120 \text{ V}}$

Answer: $I = 16.7 \text{ A}$

ANSWERS TO CHAPTER TEST IN THE INSTRUCTOR'S MANUAL

Pages 287–292

1.   a.
2.   b.
3.   Complete the chart.

Type of Power	AC Power Name	Symbol	Unit
Purely resistive	true power	P_R	watt (W)
Purely inductive	reactive power	P_Q	volt-ampere-reactive (VAR)
Purely capacitive	reactive power	P_Q	volt-ampere-reactive (VAR)
Combination R and X	apparent power	P_S	volt-ampere (VA)

4.

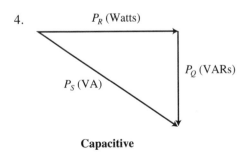

Capacitive

5.

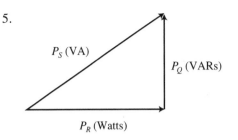

Inductive

6.   c.
7.   b.
8.   a.   Draw the equivalent circuit.

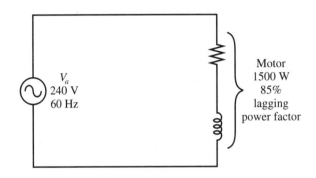

b.   Calculate the phase angle.
Formula: $\theta = \cos^{-1} \text{PF}$
Substitution: $\theta = \cos^{-1} 0.85$
Answer: $\theta = 31.8°$

c.   Calculate the reactive power.
Formula: $P_Q = P_R \times \tan \theta$
Substitution: $P_Q = 1500 \text{ W} \times \tan 31.8°$
Answer: $P_Q = 930 \text{ VAR}$

d.   Calculate the apparent power.
Formula: $P_S = \sqrt{P_R^2 + P_S^2}$
Substitution: $P_S = \sqrt{(1500 \text{ W})^2 + (930 \text{ VAR})^2}$
Answer: $P_S = 1765 \text{ VA}$

e.   Draw the power triangle.

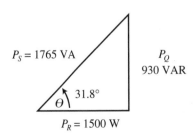

f. Calculate the current.

Formula: $I = \dfrac{P_S}{V}$

Substitution: $I = \dfrac{1765 \text{ VA}}{240 \text{ V}}$

Answer: $I = 7.35$ A

d. Calculate the value of the capacitor.

Formula: $C = \dfrac{1}{2\pi f X_C}$

Substitution: $C = \dfrac{1}{2 \times \pi \times 60 \text{ Hz} \times 61.9 \ \Omega}$

Answer: $C = 42.9 \ \mu$F

9. a. $P_Q = 930$ VAR

b. Calculate the capacitive current.

Formula: $I = \dfrac{P_Q}{V}$

Substitution: $I = \dfrac{930 \text{ VAR}}{240 \text{ V}}$

Answer: $I = 3.88$ A

c. Calculate the capacitive reactance.

Formula: $X_C = \dfrac{V}{I_C}$

Substitution: $X_C = \dfrac{240 \text{ V}}{3.88 \text{ A}}$

Answer: $X_C = 61.9 \ \Omega$

10. Formula: $I = \dfrac{P_S}{V}$

Substitution: $I = \dfrac{1500 \text{ VA}}{240 \text{ V}}$

Answer: $I = 6.25$ A

 **Chapter Test**

22

AC Power

Name: _____

Date: _____

Class: _____

Select the best answer.

_____ 1. In an ac circuit, what type of load absorbs power?
   a. Resistive.
   b. Reactive.

_____ 2. In an ac circuit, what type of load reflects power?
   a. Resistive.
   b. Reactive.

3. Complete the chart of terms related to ac power.

Type of Power	AC Power Name	Symbol	Unit
Purely resistive	_____	_____	_____
Purely inductive	_____	_____	_____
Purely capacitive	_____	_____	_____
Combination R and X	_____	_____	_____

4. In the space below, draw the phasor triangle for ac power in a circuit with resistance and capacitance. Label P_R, P_Q, P_S. Include units of measure.

5. In the space below, draw the phasor triangle for ac power in a circuit with resistance and inductance. Label P_R, P_Q, P_S. Include units of measure.

Select the best answer.

_____   6. What power factor provides maximum efficiency?
         a. 0
         b. 0.5
         c. 1.0
         d. 100

_____   7. If the power factor in a circuit is lagging, what type of reactive component has the largest value?
         a. Resistive.
         b. Inductive.
         c. Capacitive.

8. A 1500 watt motor with an 85% lagging power factor is connected to a 240 V/60 Hz supply.

     a. Draw the equivalent circuit.

     b. Calculate the phase angle.

Formula: _____

Substitution: _____

Answer: _____

c. Calculate the reactive power.

Formula: _____

Substitution: _____

Answer: _____

d. Calculate the apparent power.

Formula: _____

Substitution: _____

Answer: _____

e. Draw the power triangle.

f. Calculate the current.

Formula: _____

Substitution: _____

Answer: _____

9. For the motor circuit described in question 8, calculate the value of capacitor needed to correct to unity power factor.

a. Reactive power in the capacitor at unity power factor: _____

b. Calculate the capacitive current.

Formula: _____

Substitution: _____

Answer: _____

c. Calculate the capacitive reactance.

Formula: _____

Substitution: _____

Answer: _____

d. Calculate the value of the capacitor.

Formula: _____

Substitution: _____

Answer: _____

10. Calculate circuit current after the power factor correction of questions 8 and 9.

Formula: _____

Substitution: _____

Answer: _____

OBJECTIVES

After studying this chapter, students should be able to:
- Define resonance as it applies to ac circuits.
- Calculate the resonant frequency.
- List the characteristics of a series resonant circuit.
- Identify the typical curves for a series resonant circuit.
- Calculate values in a series resonant circuit.
- List the characteristics of a parallel resonant circuit.
- Identify the typical curves for a parallel resonant circuit.
- Calculate circuit values in a parallel resonant circuit.
- Define the bandwidth of a resonant circuit.
- Define the Q of a resonant circuit.
- Calculate the bandwidth and circuit Q.

INSTRUCTIONAL MATERIALS

Text: Pages 683–704
        Test Your Knowledge Questions, Pages 703–704
Study Guide: Pages 209–214
Laboratory Manual: Pages 237–264

ANSWERS TO TEXTBOOK

Test Your Knowledge, Pages 703–704

1. $f_r = 35.6$ Hz
2. $f_r = 50.4$ Hz
3. $f_r = 113$ H
4. $C = 338$ pF
5. $L = 0.281$ mH
6.   a. Net reactance: equals zero.
     b. Impedance: minimum.
     c. Current: maximum.
     d. Phase angle of current: in phase with voltage.
     e. Relative amount of voltage drops: rise.
     f. Power factor: unity.

7. Inductive.
8. Capacitive.
9. Minimum value.
10. Maximum value.
11.   a. $f_r = 650$ Hz
     b. $X_L = 24.5$ Ω
     c. $X_C = 24.5$ Ω
     d. $X_{net} = 0$ Ω
     e. $Z = 8$ Ω
     f. $I = 6.25$ A
     g. $V_C = 153$ V
12.   a. Net reactance: equals zero.
     b. Impedance: maximum.
     c. Current: minimum.
     d. Phase angle of current: in phase with voltage.
     e. Power factor: unity.
13. Capacitive.
14. Inductive.
15. Maximum.
16. Minimum.
17.   a. $f_r = 113$ kHz
     b. $X_L = 28.4$ Ω
     c. $X_C = 28.2$ Ω
     d. $Z_L = 10 + j28.4$ Ω (rectangular)
       $Z_L = 30.1$ Ω $\angle 70.6°$ (polar)
       $Z_C = 28.2$ Ω $\angle -90°$ (polar)
     e. $I_L = 0.83$ A $\angle -70.6°$ (polar)
       $I_L = 0.276 - j0.783$ A (rectangular)
       $I_C = 0.89$ A $\angle 90°$ (polar)
       $I_C = 0 + j0.89$ A (rectangular)
     f. $I_T = 0.276 + j0.107$ A (rectangular)
       $I_T = 0.296$ A $\angle 19.9°$ (polar)
     g. $Z = 84.5$ Ω $\angle -19.9°$ (polar)
18. $Q = 3.06$
     BW = 212 Hz
19. $Q = 2.84$
     BW = 40 kHz
20. New BW = 424 Hz
     New $Q = 1.53$
     New $r_s = 16$ Ω

ANSWERS TO STUDY GUIDE

Pages 209–214

1. f.
2. b.
3. h.
4. a.
5. e.
6. g.
7. c.
8. d.

9. Formula: $f_r = \dfrac{1}{2\pi \sqrt{LC}}$

   Substitution: $f_r = \dfrac{1}{2\pi \sqrt{0.5\ \text{H} \times 10\ \mu\text{F}}}$

   Answer: $f_r = 71.2$ Hz

10. Formula: $f_r = \dfrac{1}{2\pi \sqrt{LC}}$

    Substitution: $f_r = \dfrac{1}{2\pi \sqrt{4\ \text{mH} \times 15\ \mu\text{F}}}$

    Answer: $f_r = 650$ Hz

11. Formula: $C = \dfrac{1}{(2\pi)^2 L f_r^2}$

    Substitution: $C = \dfrac{1}{(2\pi)^2 \times 40\ \text{mH} \times (80\ \text{kHz})^2}$

    Answer: $C = 0.1$ nF

12. Formula: $L = \dfrac{1}{(2\pi)^2 C f_r^2}$

    Substitution: $L = \dfrac{1}{(2\pi)^2 \times 2\ \mu\text{F} \times (10\ \text{kHz})^2}$

    Answer: $L = 127\ \mu$H

13. a. Net reactance: Equals zero.
    b. Impedance: At a minimum.
    c. Current: At a maximum.
    d. Phase angle of current: In phase with the applied voltage.
    e. Relative amount of voltage drops: Rise.
    f. Power factor: Equals unity.
14. Inductive.
15. Capacitive.
16. Minimum value.
17. Maximum value.

18. a. Resonant frequency:

    Formula: $f_r = \dfrac{1}{2\pi \sqrt{LC}}$

    Substitution: $f_r = \dfrac{1}{2\pi \sqrt{10\ \text{mH} \times 25\ \mu\text{F}}}$

    Answer: $f_r = 318$ Hz

    b. Inductive reactance:
    Formula: $X_L = 2\pi f L$
    Substitution: $X_L = 2 \times \pi \times 318\ \text{Hz} \times 10\ \text{mH}$
    Answer: $X_L = 20\ \Omega$

    c. Capacitive reactance:

    Formula: $X_C = \dfrac{1}{2\pi f C}$

    Substitution: $X_C = \dfrac{1}{2 \times \pi \times 318\ \text{Hz} \times 25\ \mu\text{F}}$

    Answer: $X_C = 20\ \Omega$

    d. Net reactance (at resonance):
    Formula: $X_{net} = X_L - X_C$
    Substitution: $X_{net} = 20\ \Omega - 20\ \Omega$
    Answer: $X_{net} = 0$
    e. Impedance (at resonance):
    Formula: $Z = r_S$
    Substitution: $Z = 16\ \Omega$
    Answer: $Z = 16\ \Omega$
    f. Current (at resonance):
    Formula: $I = V/Z$
    Substitution: $I = 50\ \text{V} \div 16\ \Omega$
    Answer: $I = 3.13$ A
    g. Voltage across the capacitor (at resonance):
    Formula: $V_C = I \times X_C$
    Substitution: $V_C = 3.13\ \text{A} \times 20\ \Omega$
    Answer: $V_C = 62.6$ V
19. a. Net reactance: Equals zero.
    b. Impedance: At a maximum.
    c. Current: At a minimum.
    d. Phase angle of current: In phase with the applied voltage.
    e. Power factor: Equals unity.
20. Capacitive.
21. Inductive.
22. Maximum.
23. Minimum.

24. a. Resonant frequency:

Formula: $f_r = \dfrac{1}{2\pi \sqrt{LC}}$

Substitution: $f_r = \dfrac{1}{2\pi \sqrt{80\ \mu H \times 0.25\ \mu F}}$

Answer: $f_r = 35.6$ Hz

b. Inductive reactance:
Formula: $X_L = 2\pi f L$
Substitution: $X_L = 2 \times \pi \times 35.6$ kHz $\times 80\ \mu H$
Answer: $X_L = 17.9\ \Omega$

c. Capacitive reactance:

Formula: $X_C = \dfrac{1}{2\pi f C}$

Substitution:

$X_C = \dfrac{1}{2 \times \pi \times 35.6\ \text{kHz} \times 0.25\ \mu F}$

Answer: $X_C = 17.9\ \Omega$

d. Impedance of inductive branch:
Formula: $Z_L = r_S + jX_L$
Substitution: $Z_L = 10 + j17.9\ \Omega$
Answer: $Z_L = 10 + j17.9\ \Omega$ (rectangular)
$\qquad Z_L = 20.5\ \Omega$ ∠60.8° (polar)

e. Impedance of capacitive branch:
Formula: $Z_C = r_S + jX_C$
Substitution: $Z_C = 0 - j17.9\ \Omega$
Answer: $Z_C = 0 - j17.9\ \Omega$ (rectangular)
$\qquad Z_C = 17.9\ \Omega$ ∠–90° (polar)

f. Current in inductive branch:
Formula: $I_L = V/Z_L$
Substitution: $I_L = 25$ V $\div 20.5\ \Omega$ ∠60.8°
Answer: $I_L = 1.22$ A ∠–60.8° (polar)
$\qquad I_L = 0.595 - j1.06$ A (rectangular)

g. Current in capacitive branch:
Formula: $I_C = V/Z_C$
Substitution: $I_C = 25$ V $\div 17.9\ \Omega$ ∠–90°
Answer: $I_C = 1.4$ A ∠90° (polar)
$\qquad I_C = 0 + j1.4$ A (rectangular)

h. Line current:
Formula: $I_T = I_L + I_C$
Substitution:
$\qquad I_T = (0.595 - j1.06$ A$) + (0 + j1.4$ A$)$
Answer: $I_T = .595 + j0.34$ A (rectangular)
$\qquad I_T = 0.68$ A ∠30° (polar)

i. Equivalent impedance:
Formula: $Z = V/I_T$
Substitution: $Z = 25$ V $\div 0.68$ A ∠30°
Answer: $Z = 36.8\ \Omega$ ∠–30°

25. Formulas: $Q = X_L/r_S$ and   BW $= f_r/Q$
Substitution: $Q = 20\ \Omega \div 16\ \Omega$
Answer: $Q = 1.25$
Substitution: BW $= 318$ Hz $\div 1.25$
BW $= 254$ Hz

26. Formulas: $Q = X_L/r_S$ and   BW $= f_r/Q$
Substitution: $Q = 17.9\ \Omega \div 10\ \Omega$
Answer: $Q = 1.79$
Substitution: BW $= 35.6$ Hz $\div 1.79$
BW $= 19.9$ Hz

27. New BW $= 39.8$ Hz
Formulas: $Q = f_r/$BW   and $r_S = X_L/Q$
Substitution: $Q = 35.6 \div 39.8$
Answer: $Q = 0.894$
Substitution: $r_S = 17.9\ \Omega \div 0.894$
Answer: $r_S = 22.3\ \Omega$

ANSWERS TO CHAPTER TEST IN THE INSTRUCTOR'S MANUAL

Pages 299–308

1. Resonance
2. bell curve
3. tank circuit
4. bandwidth
5. volts (or microvolts)
6. Q

7. Formula: $f_r = \dfrac{0.159}{\sqrt{LC}}$

Substitution: $f_r = \dfrac{0.159}{\sqrt{0.25\ \text{H} \times 50\ \mu F}}$

Answer: $f_r = 45$ Hz

8. Formula: $f_r = \dfrac{0.159}{\sqrt{LC}}$

Substitution: $f_r = \dfrac{0.159}{\sqrt{8\ \text{mH} \times 10\ \mu F}}$

Answer: $f_r = 562$ Hz

9. below resonance: C
above resonance: L

10. below resonance: L
    above resonance: C

11. a. Impedance: minimum
    b. Current: maximum

12. a. Impedance: maximum
    b. Current: minimum

13. Formula: $f_r = \dfrac{0.159}{\sqrt{LC}}$

    Substitution: $f_r = \dfrac{0.159}{\sqrt{50 \text{ mH} \times 15 \text{ μF}}}$

    Answer: $f_r = 184$ Hz

14. a. Inductive reactance:
    Formula: $X_L = 2\pi f L$
    Substitution: $X_L = 2 \times \pi \times 184 \text{ Hz} \times 50 \text{ mH}$
    Answer: $X_L = 57.8$ Ω

    b. Capacitive reactance:

    Formula: $X_C = \dfrac{1}{2\pi f C}$

    Substitution: $X_C = \dfrac{1}{2 \times \pi \times 184 \text{ Hz} \times 15 \text{ μF}}$

    Answer: $X_C = 57.7$ Ω

    c. $X_{net} = 0$ Ω (cancel at resonance)

    d. $Z = r_S = 8$ Ω

15. Formula: $I = \dfrac{V}{Z}$

    Substitution: $I = \dfrac{30 \text{ V}}{8 \text{ Ω}}$

    Answer: $I = 3.75$ A

16. Formula: $f_r = \dfrac{0.159}{\sqrt{LC}}$

    Substitution: $f_r = \dfrac{0.159}{\sqrt{90 \text{ μH} \times 0.3 \text{ μF}}}$

    Answer: $f_r = 30.6$ kHz

17. a. Inductive reactance:
    Formula: $X_L = 2\pi f L$
    Substitution: $X_L = 2 \times \pi \times 30.6 \text{ kHz} \times 90 \text{ μH}$
    Answer: $X_L = 17.3$ Ω

    b. Capacitive reactance:

    Formula: $X_C = \dfrac{1}{2\pi f C}$

    Substitution:

    $X_C = \dfrac{1}{2 \times \pi \times 30.6 \text{ kHz} \times 0.3 \text{ μF}}$

    Answer: $X_C = 17.3$ Ω

18. a. Formula: $Z_L = r_S + jX_L$
    Answer (rectangular): $Z_L = 12 \text{ Ω} + j17.3 \text{ Ω}$
    Answer (polar): $Z_L = 21.1 \text{ Ω} \angle 55.3°$

    b. Formula: $Z_C = R - jX_C$
    Answer (rectangular): $Z_C = 0 \text{ Ω} - j17.3 \text{ Ω}$
    Answer (polar): $Z_C = 17.3 \text{ Ω} \angle{-90°}$

19. a. Current in the inductance branch:

    Formula: $I_L = \dfrac{V}{Z_L}$

    Substitution: $I_L = \dfrac{10 \text{ V} \angle 0°}{21.1 \text{ Ω} \angle 55.3°}$

    Answer (polar): $I_L = 0.474 \text{ A} \angle{-55.3°}$

    Answer (rectangular): $I_L = 0.270 - j0.390$ A

    b. Current in the capacitance branch:

    Formula: $I_C = \dfrac{V}{Z_C}$

    Substitution: $I_C = \dfrac{10 \text{ V} \angle 0°}{17.3 \text{ Ω} \angle{-90°}}$

    Answer (polar): $I_C = 0.578 \text{ A} \angle 90°$

    Answer (rectangular): $I_C = 0 + j0.578$ A

    c. Formula: $I_T = I_L + I_C$
    Substitution:
    $I_T = (0.270 - j0.390 \text{ A}) + (0 + j0.578 \text{ A})$
    Answer (rectangular): $I_T = 0.270 + j0.188$ A
    Answer (polar): $I_T = 0.329 \text{ A} \angle 34.8°$

20. a.   Circuit Q:

Formula: $Q = \dfrac{X_L}{r_S}$

Substitution: $Q = \dfrac{17.3\ \Omega}{12\ \Omega}$

Answer: $Q = 1.44$

b.   Bandwidth:

Formula: $BW = \dfrac{f_r}{Q}$

Substitution: $BW = \dfrac{30.6\ kHz}{1.44}$

Answer: $BW = 21.3\ kHz$

Chapter Test

23

Name: _____

Date: _____

Class: _____

Resonance

For questions 1 through 6, fill in the blanks with technical terms from the chapter.

1. _____ occurs when capacitive reactance equals inductive reactance in an ac circuit.

2. The shape of a typical curve showing the response of a resonant circuit is called a _____

   _____.

3. A parallel resonant circuit is also called a _____ _____.

4. The _____ is the name of the range of frequencies between the half-power points.

5. Selectivity is the strength of the signal, measured in _____, within the accepted bandwidth.

6. The circuit factor _____ is a number with no units. It gives the ratio of inductive reactance

   to the circuit resistance.

7. Calculate the resonant frequency of a circuit containing 0.25 H inductance and 50 μF capacitance.

Formula: _____

Substitution: _____

Answer: _____

8. Calculate the resonant frequency of a circuit with 10 μF capacitance and 8 mH inductance.

Formula: _____

Substitution: _____

Answer: _____

For questions 9 and 10, circle the L or the C to indicate if the net reactance is inductive or capacitive under given conditions.

9. Describe net reactance in a typical series resonant circuit under these conditions:

	below resonance	at resonance	above resonance
net reactance	L or C	_____	L or C

10. Describe net reactance in a typical parallel resonant circuit under these conditions:

	below resonance	at resonance	above resonance
net reactance	L or C	_____	L or C

11. Answer *maximum* or *minimum* for the performance characteristic of a typical series circuit at resonance.

   a. Impedance: (maximum/minimum) _____

   b. Current: (maximum/minimum) _____

12. Answer *maximum* or *minimum* for the performance characteristic of a typical parallel circuit at resonance.

   a. Impedance: (maximum/minimum) _____

   b. Current: (maximum/minimum) _____

For questions 13 though 15, a 50 mH inductor is connected in series with a 15 μF capacitor and a 30 volt generator. The effective r_S in the circuit is 8 ohms. Find the requested values.

13. Find the resonant frequency.

Formula: _____

Substitution: _____

Answer: _____

14. Find the reactances and impedance at resonance.

a. Inductive reactance:

Formula: _____

Substitution: _____

Answer: _____

b. Capacitive reactance:

Formula: _____

Substitution: _____

Answer: _____

c. Net reactance:

Formula: _____

Substitution: _____

Answer: _____

Name: ______________________________

d. Impedance:

Formula: ____

Substitution: ____

Answer: ____

15. Find the circuit current at resonance.

Formula: ____

Substitution: ____

Answer: ____

For questions 16 through 20, a signal generator with 10 volts is applied to the parallel combination of a 0.3 µF capacitor and 90 µH inductor. The effective r_S of the inductor is 12 ohms. Find the requested values.

16. Find the resonant frequency.

Formula: _____

Substitution: _____

Answer: _____

17. Find the reactances at resonance.

a. Inductive reactance:

Formula: _____

Substitution: _____

Answer: _____

b. Capacitive reactance:

Formula: _____

Substitution: _____

Answer: _____

18. Find the impedances at resonance.

   a. Impedance of inductance branch:

Formula: _____

Answer (rectangular): _____

Answer (polar): _____

b. Impedance of capacitance branch:

Formula: _____

Answer (rectangular): _____

Answer (polar): _____

19. Find the currents at resonance.

a. Current in the inductance branch:

Formula: _____

Substitution: _____

Answer (polar): _____

Answer (rectangular): _____

b. Current in the capacitance branch:

Formula: _____

Substitution: _____

Answer (polar): _____

Answer (rectangular): _____

c. Line current:

Formula: _____

Substitution: _____

Answer (rectangular): _____

Answer (polar): _____

20. Calculate the bandwidth and circuit Q.

    a. Circuit Q:

    Formula: _____

    Substitution: _____

    Answer: _____

    b. Bandwidth:

    Formula: _____

    Substitution: _____

    Answer: _____

Chapter 24

Filter Circuits

OBJECTIVES

After studying this chapter, students should be able to:
- Describe how to remove a dc voltage from an ac signal.
- Describe how to separate ac and dc voltages.
- Describe how a filter circuit separates different ac frequencies.
- Describe how a low pass filter removes high frequencies.
- Calculate the cutoff frequency and output voltage when a low pass filter is used as a voltage divider.
- Describe how a high pass filter removes low frequencies.
- Calculate the cutoff frequency and output voltage when a high pass filter is used as a voltage divider.
- Describe how a resonant circuit is used as a band pass filter.
- Describe how a resonant circuit is used as a band stop filter.

INSTRUCTIONAL MATERIALS

Text: Pages 705–734
        Test Your Knowledge Questions, Pages 731–733
Study Guide: Pages 215–220
Laboratory Manual: Pages 265–275

ANSWERS TO TEXTBOOK

Test Your Knowledge, Pages 731–733

1. Remove a dc voltage from an ac signal, remove an ac ripple from a dc voltage, separate ac and dc for use in a circuit.

2. A capacitor charges to the dc voltage and stops any further dc from passing. The changing ac voltages pass through with only the reactance of the capacitor.

3. A transformer passes only a changing signal from the primary to secondary.

4. A bypass capacitor provides a parallel path for the ac voltages, while not affecting the dc voltages.

5. Inductor.

6. Capacitor.

7. a. Inductor.
   b. Capacitor.

8. Refer to figure 24-8.

9. At 1000 Hz:
   $X_L = 628\ \Omega$
   $X_C = 31.8\ \Omega$
   At 50 kHz:
   $X_L = 31.4\ k\Omega$
   $X_C = 0.637\ \Omega$

10. Use the value that makes the reactance equal the resistance.

11. $f = 3.18$ kHz
    $V = 42.4$ V

12. At 50 Hz:
    $X_L = 15.7\ \Omega$
    $X_C = 796\ \Omega$
    At 20 kHz:
    $X_L = 6280\ \Omega$
    $X_C = 2\ \Omega$

13. Use the value that makes the reactance equal the resistance.

14. $f = 79.6$ kHz
    $V = 7.07$ V

15. For band pass, the series resonant circuit is placed in series with the load, because it offers minimum impedance at the resonant frequency.

16. For band pass, the parallel resonant circuit is placed in parallel with the load, because it offers maximum impedance at the resonant frequency.

17. For band stop, the series resonant circuit is placed in parallel with the load. The minimum impedance will pass frequencies near resonance to ground.

16. For band stop, the parallel resonant circuit is placed in series with the load. The maximum impedance stops frequencies near resonance.

ANSWERS TO STUDY GUIDE

Pages 215–220

1. a and e.
2. h.
3. b.
4. d, f, and g.
5. c.
6. a.
7. b.
8. d.
9. a.
10. b.
11. a.
12. b.
13. Refer to figure 24-8.
14. a. X_L at 2000 Hz:
    Formula: $X_L = 2\pi fL$
    Substitution: $X_L = 2\pi \times 2000$ Hz $\times 100$ mH
    Answer: $X_L = 1257\ \Omega$
    b. X_L at 75 kHz:
    Formula: $X_L = 2\pi fL$
    Substitution: $X_L = 2\pi \times 75$ kHz $\times 100$ mH
    Answer: $X_L = 47.1\ \Omega$
    c. X_C at 2000 Hz:
    Formula: $X_C = 1 \div 2\pi fC$
    Substitution:
    $X_C = 1 \div (2\pi \times 2000$ Hz $\times 5\ \mu F)$
    Answer: $X_C = 15.9\ \Omega$
    d. X_C at 75 kHz:
    Formula: $X_C = 1 \div 2\pi fC$
    Substitution: $X_C = 1 \div (2\pi \times 75$ kHz $\times 5\ \mu F)$
    Answer: $X_C = 0.424\ \Omega$
15. c.
16. a. Formula: $f = 1 \div 2\pi X_C C$
    Substitution: $f = 1 \div (2\pi \times 20\ \Omega \times 10\ \mu F)$
    Answer: $f = 796$ Hz
    b. Formula: $V = V_{in} \times 0.707$
    Substitution: $V = 50$ V $\times 0.707$
    Answer: $V = 35.4$ V

17. a. X_L at 120 Hz:
    Formula: $X_L = 2\pi fL$
    Substitution: $X_L = 2\pi \times 120$ Hz $\times 15$ mH
    Answer: $X_L = 11.3\ \Omega$
    b. X_L at 15 kHz:
    Formula: $X_L = 2\pi fL$
    Substitution: $X_L = 2\pi \times 15$ kHz $\times 100$ mH
    Answer: $X_L = 1.4$ kΩ
    c. X_C at 120 Hz:
    Formula: $X_C = 1 \div 2\pi fC$
    Substitution: $X_C = 1 \div (2\pi \times 120$ Hz $\times 2\ \mu F)$
    Answer: $X_C = 663\ \Omega$
    d. X_C at 15 kHz:
    Formula: $X_C = 1 \div 2\pi fC$
    Substitution: $X_C = 1 \div (2\pi \times 15$ kHz $\times 2\ \mu F)$
    Answer: $X_C = 5.3\ \Omega$
18. c.
19. a. Formula: $f = 1 \div 2\pi X_C C$
    Substitution: $f = 1 \div (2\pi \times 50\ \Omega \times 0.1\ \mu F)$
    Answer: $f = 31.8$ kHz
    b. Formula: $V = V_{in} \times 0.707$
    Substitution: $V = 20$ V $\times 0.707$
    Answer: $V = 14.1$ V
20. b. Series resonant band pass.
    c. Series resonant band stop.
    d. Parallel resonant band pass.
    a. Parallel resonant band stop.

ANSWERS TO CHAPTER TEST IN THE INSTRUCTOR'S MANUAL

Pages 313–322

1. dc, ac
2. saturated
3. bypass
4. -3db point, cutoff frequency, half-power point (in any order)
5. a.
6. b.
7. d.
8. a.
9. b.
10. a. inductor
    b. capacitor
11. a. X_L at 5 kHz:
    Formula: $X_L = 2\pi fL$
    Substitution: $X_L = 2 \times \pi \times 5$ kHz $\times 3$ mH
    Answer: $X_L = 94.2\ \Omega$

b. X_C at 5 kHz:

Formula: $X_C = \dfrac{1}{2\pi fC}$

Substitution: $X_C = \dfrac{1}{2 \times \pi \times 5\ \text{kHz} \times 0.01\ \mu\text{F}}$

Answer: $X_C = 3.18\ \text{k}\Omega$

12. a. X_L at 150 kHz:
    Formula: $X_L = 2\pi fL$
    Substitution: $X_L = 2 \times \pi \times 150\ \text{kHz} \times 3\ \text{mH}$
    Answer: $X_L = 2.83\ \text{k}\Omega$

b. X_C at 150 kHz:

Formula: $X_C = \dfrac{1}{2\pi fC}$

Substitution:

$$X_C = \dfrac{1}{2 \times \pi \times 150\ \text{kHz} \times 0.01\ \mu\text{F}}$$

Answer: $X_C = 106\ \Omega$

13. Formula: $f = \dfrac{1}{2\pi CX_C}$

Substitution: $f = \dfrac{1}{2 \times \pi \times 5\ \mu\text{F} \times 30\ \Omega}$

Answer: $f = 1.06\ \text{kHz}$

14. Formula: $V_{out} = V_{in} \times \dfrac{X_C}{\sqrt{R^2 + X_C^{\,2}}}$

Substitution: $V_{out} = \dfrac{100\ \text{V} \times 30\ \Omega}{\sqrt{(30\ \Omega)^2 + (30\ \Omega)^2}}$

Answer: $V_{out} = 70.7\ \text{V}$

Note: V_{out} = 70.7% of applied voltage.

15. a. X_L at 60 Hz:
    Formula: $X_L = 2\pi fL$
    Substitution: $X_L = 2 \times \pi \times 60\ \text{Hz} \times 20\ \text{mH}$
    Answer: $X_L = 7.54\ \Omega$

b. X_C at 60 Hz:

Formula: $X_C = \dfrac{1}{2\pi fC}$

Substitution: $X_C = \dfrac{1}{2 \times \pi \times 60\ \text{Hz} \times 1.8\ \mu\text{F}}$

Answer: $X_C = 1.47\ \text{k}\Omega$

16. a. X_L at 10 kHz:
    Formula: $X_L = 2\pi fL$
    Substitution: $X_L = 2 \times \pi \times 10\ \text{kHz} \times 20\ \text{mH}$
    Answer: $X_L = 1.26\ \text{k}\Omega$

b. X_C at 10 kHz:

Formula: $X_C = \dfrac{1}{2\pi fC}$

Substitution: $X_C = \dfrac{1}{2 \times \pi \times 10\ \text{kHz} \times 1.8\ \mu\text{F}}$

Answer: $X_C = 8.84\ \Omega$

17. c.

18. Formula: $f = \dfrac{1}{2\pi CX_C}$

Substitution: $f = \dfrac{1}{2 \times \pi \times 0.5\ \mu\text{F} \times 80\ \Omega}$

Answer: $f = 3.98\ \text{kHz}$

19. Formula: $V_{out} = V_{in} \times \dfrac{X_C}{\sqrt{R^2 + X_C^{\,2}}}$

Substitution: $V_{out} = 15\ \text{V} \times \dfrac{80\ \Omega}{\sqrt{(80\ \Omega)^2 + (80\ \Omega)^2}}$

Answer: $V_{out} = 10.6\ \text{V}$

Note: V_{out} = 70.7% of applied voltage.

20. a. Series resonant band pass.
    d. Series resonant band stop.
    c. Parallel resonant band pass.
    b. Parallel resonant band stop.

Name: _____

Date: _____

Class: _____

Filter Circuits

Fill in the blanks.

1. The purpose of a blocking capacitor is to stop _____ voltages, while passing

   _____ voltages. (Answer with ac or dc.)

2. When an applied dc voltage is large enough to produce maximum magnetism in a transformer, the transformer

   becomes _____.

3. A parallel path for the ac signal around the load resistor is provided by a _____ capacitor.

4. List the three names for the point on a response curve where the frequency is equal to 70.7% of maximum.

   _____

   _____

   _____

Select the best answer.

_____ 5. A capacitor removes a dc voltage from an ac signal:
   a. by charging to the dc voltage.
   b. by passing the dc voltage to ground.
   c. by passing the ac signal to ground.
   d. by offering near zero reactance to dc voltages.

_____ 6. A transformer removes a dc voltage from an ac signal by:
   a. charging to the dc voltage.
   b. passing only changing voltages to the secondary.
   c. passing the ac signal to ground.
   d. providing a very high secondary resistance to dc voltage.

_____ 7. A capacitor is used to provide a separate path for an ac signal from a dc voltage by:
   a. shorting the dc voltage to ground.
   b. offering a very high impedance to ac signals.
   c. offering a low impedance path for dc voltage.
   d. offering a low impedance path for ac around a resistor.

_____   8. Which component offers the least opposition to low frequencies?
         a. Inductor.
         b. Capacitor.

_____   9. Which component offers the least opposition to high frequencies?
         a. Inductor.
         b. Capacitor.

10. Answer with *inductor* or *capacitor*. If an inductor and capacitor are connected in parallel as a filter circuit, which path is taken by:

    a. dc voltages and low frequency ac voltages? _____

    b. high frequency ac signals? _____

11. An audio signal of 5 kHz is transmitted with a 150 kHz radio frequency carrier. The audio signal is removed using a low pass filter ($L = 3$ mH and $C = 0.01$ μF). Calculate X_L and X_C at the audio frequency.

    a. X_L at 5 kHz:

        Formula: _____

        Substitution: _____

        Answer: _____

    b. X_C at 5 kHz:

        Formula: _____

        Substitution: _____

        Answer: _____

12. Calculate X_L and X_C at the radio frequency for the filter in question 11.

    a. X_L at 150 kHz:

        Formula: _____

        Substitution: _____

        Answer: _____

    b. X_C at 150 kHz:

        Formula: _____

        Substitution: _____

        Answer: _____

13. Determine the cutoff frequency of the low pass filter in the following figure.

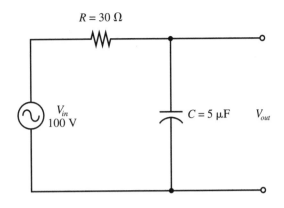

Formula: _____

Substitution: _____

Answer: _____

14. Determine the output voltage for the circuit in question 13.

Formula: _____

Substitution: _____

Answer: _____

15. A 60 Hz hum is mixed with an audio signal in an amplifier. The hum is removed using the 10 kHz high pass filter shown in the figure below. Calculate X_L and X_C at the hum frequency.

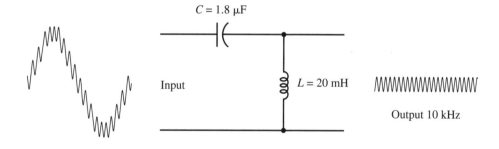

60 Hz with 10 kHz

a. X_L at 60 Hz:

Formula: _____

Substitution: _____

Answer: _____

b. X_C at 60 Hz:

Formula: _____

Substitution: _____

Answer: _____

16. Calculate X_L and X_C for the high pass filter of question 15.

　　a.　X_L at 10 kHz:

　　　　Formula: _____

　　　　Substitution: _____

　　　　Answer: _____

　　b.　X_C at 10 kHz:

　　　　Formula: _____

　　　　Substitution: _____

　　　　Answer: _____

_____ 17. In a high pass reactive voltage divider, the reactance should be:
      a. greater than the value of the resistor.
      b. less than the value of the resistor.
      c. equal to the value of the resistor.

18. Determine the cutoff frequency of the high pass filter in the following figure.

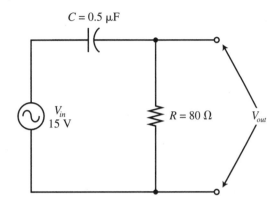

Formula: _____

Substitution: _____

Answer: _____

19. Determine the output voltage of the high pass filter in question 18.

Formula: _____

Substitution: _____

Answer: _____

20. Identify the schematics of the resonant filters shown using the following names.

Series resonant band pass. _____

Series resonant band stop. _____

Parallel resonant band pass. _____

Parallel resonant band stop. _____

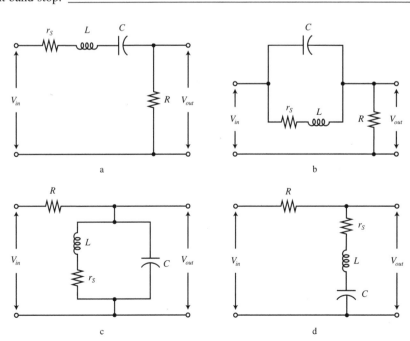

Single- and Polyphase Systems

OBJECTIVES

After studying this chapter, students should be able to:

- Describe how a single-phase transformer can produce a three-wire single-phase output.
- Describe the typical three-wire single-phase electrical distribution system used in residential applications.
- Describe a typical two-phase electrical system.
- Calculate total current in a two-phase system.
- Describe the general principles of a three-phase system.
- Draw the connections for a four-wire wye connected three-phase system.
- Calculate the voltage at the output of a wye connected system.
- Draw the connections for a three-wire delta connected three-phase system.
- Calculate the voltage at the output of a delta connected system.
- Draw the generator to output combinations of wye and delta connections.

INSTRUCTIONAL MATERIALS

Text: Pages 735–748
     Test Your Knowledge Questions, Page 747
Study Guide: Pages 221–224

ANSWERS TO TEXTBOOK

Test Your Knowledge, Page 747

1. A center-tapped transformer splits the ac into two voltages 180° apart. The third wire is ground.
2. Power from the power company comes into the circuit breaker panel as a three-wire single-phase system. There is 120 volts on two wires. The third wire is the ground wire.

3. Two independent sine waves travel at 90° apart. Two phase systems come in three- or four-wire configurations.
4. $I_A = 2$ A $\angle 90°$ (polar)
   $I_A = 0 + j2$ A (rectangular)
   $I_B = 2$ A $\angle 0°$ (polar)
   $I_B = 2 + j0$ A (rectangular)
   $I_T = 2 + j2$ A (rectangular)
   $I_T = 2.83$ $\angle 45°$ (polar)
5. The three-phase system has three coils spaced 120° apart. These produce three sine waves, each with a phase difference of 120°. Three-phase systems come in three- and four-wire configurations.
6. Refer to figure 25-9.
7. 173 V
8. Refer to figure 25-10.
9. 34.6 A
10. Refer to figures 25-11 through 25-14.

ANSWERS TO STUDY GUIDE

Pages 221–224

1. a.
2. c.
3. c.
4. Formulas: $I_T = I_A + I_B$ and $I = V/R$
   Substitution: $I_A = 240$ A $\angle 90°$ ÷ 80 Ω
   Answer: $I_A = 3$ A $\angle 90°$ (polar)
      $I_A = 0 + j3$ A (rectangular)
   Substitution: $I_B = 240$ A $\angle 0°$ ÷ 80 Ω
   Answer: $I_B = 3$ A $\angle 0°$ (polar)
      $I_B = 3 + j0$ A (rectangular)
   Substitution: $I_T = (0 + j3$ A) + (3 + j0 A)
   Answer: $I_T = 3 + j3$ A (rectangular)
      $I_T = 4.24$ A $\angle 45°$
5. b.
6. Refer to textbook figure 25-9.

7. Formula: $V_{out} = \sqrt{3}V_{\phi}$

   Substitution: $V_{out} = \sqrt{3} \times 220$ V

   Answer: $V_{out} = 381$ V

8. Refer to textbook figure 25-10.

9. Formula: $I_{out} = \sqrt{3}I_{\phi}$

   Substitution: $I_{out} = \sqrt{3} \times 30$ A

   Answer: $I_{out} = 52$ A

10. Refer to textbook figure 25-11.
11. Refer to textbook figure 25-12.
12. Refer to textbook figure 25-13.
13. Refer to textbook figure 25-14.

ANSWERS TO CHAPTER TEST IN THE INSTRUCTOR'S MANUAL

Pages 325–329

1. a.
2. c.
3. b.
4. c.

5. a. Current of phase A:

   Formula: $I_A = \dfrac{V}{R}$

   Substitution: $I_A = \dfrac{240 \text{ V } \angle 90°}{40 \text{ }\Omega}$

   Answer: $I_A = 6$ A $\angle 90°$

   b. Current of phase B:

   Formula: $I_B = \dfrac{V}{R}$

   Substitution: $I_B = \dfrac{240 \text{ V } \angle 0°}{40 \text{ }\Omega}$

   Answer: $I_B = 6$ A $\angle 0°$

   c. Formula: $I_T = I_A + I_B$
   Substitution: $I_T = 6$ A $\angle 90° + 6$ A $\angle 0°$
   Intermediate step: $I_T = \sqrt{(6 \text{ A})^2 + (6 \text{ A})^2}$
   Answer: $I_T = 8.49$ A $\angle 45°$

6.

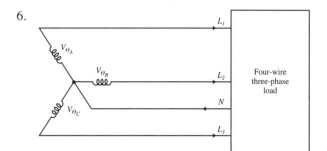

7. Formula: $V_L = \sqrt{3} \ V_{\theta}$

   Substitution: $V_L = \sqrt{3} \times 60$ V

   Answer: $V_L = 104$ V

8.

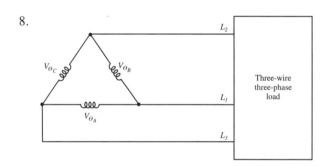

9. Formula: $I_L = \sqrt{3} \ I_{\theta}$

   Substitution: $I_L = \sqrt{3} \times 25$ A

   Answer: $I_L = 43.3$ A

10.

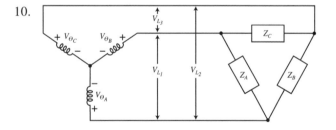

Chapter Test

25

Name: _____

Date: _____

Class: _____

Single- and Polyphase Systems

Select the best answer.

_____ 1. A single-phase transformer can produce a three-wire single-phase output by splitting it into two single phase voltages:
   a. 180° apart.
   b. 120° apart.
   c. 90° apart.
   d. 0° apart.

_____ 2. In a typical two-phase electrical system, each phase is:
   a. 180° apart.
   b. 120° apart.
   c. 90° apart.
   d. 0° apart.

_____ 3. In a typical three-phase electrical system, each phase is:
   a. 180° apart.
   b. 120° apart.
   c. 90° apart.
   d. 0° apart.

_____ 4. The typical three-wire single-phase electrical distribution system used in residential applications has:
   a. three 120 volt lines.
   b. two 240 volt lines with a neutral.
   c. two 120 volt lines with a neutral.
   d. three 240 volt lines.

5. Calculate the total current in a two-phase system with a load of 40 ohms on each phase. Both phases have an output voltage of 240 volts. Phase A has a phase shift of 90°, and phase B has a phase shift of 0°.

a. Current of phase A:

Formula: _____

Substitution: _____

Answer: _____

b. Current of phase B:

Formula: _____

Substitution: _____

Answer: _____

c. Total current:

Formula: _____

Substitution: _____

Intermediate step: _____

Answer: _____

6. Use the space below to draw the connections for a four-wire wye connected three-phase system.

7. Calculate the voltage at the output of a wye connected system if each phase produces 60 volts.

Formula: _____

Substitution: _____

Answer: _____

8. Use the space below to draw the connections for a three-wire delta connected three-phase system.

9. Calculate the current at the output of a delta connected system supplying a balanced load with 25 amps in each phase.

Formula: _____

Substitution: _____

Answer: _____

10. Use the space below to draw the windings of a wye generator to a delta output.

Diodes and Power Supplies

OBJECTIVES

After studying this chapter, students should be able to:
- Define technical terms related to semiconductors.
- Identify the characteristics of P-type and N-type semiconductor materials.
- Describe the current in a semiconductor.
- Use a series of diagrams to demonstrate the operation of a junction diode.
- Describe typical applications for diodes.
- Identify how diodes are rated.
- Test a diode with an ohmmeter.
- Describe the operation of various types of diode circuits.
- Describe the operation of rectifier circuits.
- Calculate the dc output voltage of a rectifier circuit.
- Describe the operation of a simple capacitive power supply filter.
- Calculate the output voltage of a power supply with a capacitive filter.
- Calculate output voltage ripple.
- Recognize the schematic diagrams of voltage multipliers.
- Describe the function and operation of a zener diode.
- Calculate the circuit values of a zener regulator.
- Identify other zener diode circuits.

INSTRUCTIONAL MATERIALS

Text: Pages 749–790
　　　　Test Your Knowledge Questions, Pages 788–789
Study Guide: Pages 225–237
Laboratory Manual: Pages 277–307

ANSWERS TO TEXTBOOK

Test Your Knowledge, Pages 788–789
1. Rectifiers, voltage limiters, voltage regulators.
2. Pentavalent.
3. Trivalent.
4. Holes.
5. Electrons.
6. Electrons.
7. Holes.
8. Refer to figures 26-6, 26-7, 26-9.
9. The area near the junction is void of mobile carriers.
10. The depletion region collapses.
11. The depletion region expands.
12. 0.2 V for Ge and 0.6 for Si.
13. The full applied voltage.
14. The current is limited only by the external circuit.
15. The current is near zero.
16. Rectifier, one-way switch, voltage reference.
17. Reverse breakdown voltage, forward current.
18. A diode conducts in only one direction.
19. Low resistance.
20. Infinity.
21. In forward bias, current is allowed to flow. In reverse bias, the applied voltage is blocked.
22. Refer to figure 26-13.
23. Makes a sine wave look square. Clips off the peaks of the wave. See figures 26-18 and 26-19.
24. It changes the wave's centerline.
25. Refer to figures 26-21 and 26-22.
26. The rectifier conducts one half cycle and blocks the other half cycle.
27. V_{dc} = 10.8 V
28. Refer to figure 26-27.
29. One diode conducts the wave through the load for one half cycle. A second diode conducts through the load on the other half cycle.
30. V_{dc} = 21.6 V
31. Refer to figure 26-29.
32. Two diodes conduct the wave through the load on one half cycle. Two other diodes direct the current through the load for the other half cycle.
33. V_{dc} = 27 V
34. Refer to figure 26-38.

35. $V_{dc} = 48$ V

36. Ripple = 0.56%

37. Refer to figures 26-40 through 26-43.

38. Refer to figure 26-44.

39. $V_Z = 12$ V

   $I_{Z(max)} = 275$ mA

   $P_Z = 3.3$ W

   $I_{R_S} = 275$ mA

   $V_{R_S} = 8$ V

   $R_S = 29$ Ω

   $P_{R_S} = 2.2$ W

40. Refer to figures 26-46 and 26-47.

ANSWERS TO STUDY GUIDE

Pages 225–237

1. w.
2. t.
3. z.
4. j.
5. u.
6. s.
7. y.
8. h.
9. k.
10. o.
11. n.
12. q.
13. r.
14. v.
15. m.
16. p.
17. i.
18. l.
19. c.
20. b.
21. e.
22. a.
23. g.
24. f.
25. x.
26. d.
27. aa.
28. c.
29. a.
30. b.
31. a.

32. a. Zero.
    b. None.
    c. None.
    d. Normal.
    e. Zero.

33. a. Maximum.
    b. Majority.
    c. Forward.
    d. Collapsed.
    e. 0.2 to 0.6.

34. a. Minimum.
    b. Minority.
    c. Reverse.
    d. Expanded.
    e. Applied voltage.

35. a.

36. d.

37. Refer to textbook figure 26-13.

38. Refer to textbook figure 26-14. Low resistance.

39. Refer to textbook figure 26-13.

40. Refer to textbook figure 26-14. High resistance.

41. Refer to textbook figure 26-15.

42. Refer to textbook figure 26-17.

43. Refer to textbook figure 26-19.

44. Refer to textbook figure 26-21.

45. Refer to textbook figure 26-22.

46. Formulas:
    peak = $V_{rms} \div 0.707$ and V_{dc} = peak × 0.318
    Substitution: peak = 48 V ÷ 0.707
    Answer: peak = 67.9 V
    Substitution: V_{dc} = 67.9 V × 0.318
    Answer: V_{dc} = 21.6 V

47. Refer to textbook figures 26-27 and 26-28.

48. Refer to textbook figures 26-27 and 26-28.

49. Formulas:
    peak = $V_{rms} \div 0.707$ and V_{dc} = peak × 0.636
    Substitution: peak = 48 V ÷ 0.707
    Answer: peak = 67.9 V
    Substitution: V_{dc} = 67.9 V × 0.636
    Answer: V_{dc} = 43.2 V

50. Refer to textbook figure 26-29.

51. Refer to textbook figure 26-29.

52. Formulas:
    peak = $V_{rms} \div 0.707$ and V_{dc} = peak × 0.636
    Substitution: peak = 48 V ÷ 0.707
    Answer: peak = 67.9 V
    Substitution: V_{dc} = 67.9 V × 0.636
    Answer: V_{dc} = 43.2 V

53. Refer to textbook figure 26-38.

54. Formula: V_{dc} = peak voltage (no-load)
    Substitution/Answer: V_{dc} = 48 V

55. Formula: $\% \text{ ripple} = \dfrac{ac}{dc}$

    Substitution: $\% \text{ ripple} = \dfrac{200 \text{ mV}}{50 \text{ V}}$

    Answer: $\% \text{ ripple} = 0.4\%$

56. Refer to textbook figure 26-42.
57. Refer to textbook figure 26-41.
58. Refer to textbook figure 26-40.
59. Refer to textbook figure 26-43.
60. Refer to textbook figure 26-44.

61. a. Zener voltage:
    Formula: $V_Z = V_L$
    Substitution/Answer: $V_Z = 6 \text{ V}$

    b. Maximum zener current:
    Formula: $I_{Z(max)} = I_L + 10\%$
    Substitution: $I_{Z(max)} = 500 \text{ mA} + 50 \text{ mA}$
    Answer: $I_{Z(max)} = 550 \text{ mA}$

    c. Zener power rating:
    Formula: $P_Z = I_{Z(max)} \times V_Z$
    Substitution: $P_Z = 550 \text{ mA} \times 6 \text{ V}$
    Answer: $P_Z = 3.3 \text{ W}$

    d. Current through the series resistor:
    Formula: $I_{R_S} = I_{Z(max)}$
    Substitution/Answer: $I_{R_S} = 550 \text{ mA}$

    e. Voltage across the series resistor:
    Formula: $V_{R_S} = V_{dc} - V_Z$
    Substitution: $V_{R_S} = 12 \text{ V} - 6 \text{ V}$
    Answer: $V_{R_S} = 6 \text{ V}$

    f. Ohmic value of the series resistor:

    Formula: $R_S = \dfrac{V_{R_S}}{I_{R_S}}$

    Substitution: $R_S = \dfrac{6 \text{ V}}{550 \text{ mA}}$

    Answer: $R_S = 10.9 \ \Omega$

    g. Power dissipated across the series resistor:
    Formula: $P_{R_S} = V_{R_S} \times I_{R_S}$
    Substitution: $P_{R_S} = 6 \text{ V} \times 550 \text{ mA}$
    Answer: $P_{R_S} = 3.3 \text{ W}$

62. Refer to textbook figure 26-45.
63. Refer to textbook figure 26-47.
64. Refer to textbook figure 26-46.

ANSWERS TO CHAPTER TEST IN THE INSTRUCTOR'S MANUAL

Pages 337–343

1. rectification
2. trivalent
3. pentavalent
4. covalent
5. depletion
6. Forward
7. barrier
8. clipper
9. clamper
10. reverse
11. d.

Complete the following chart describing semiconductors.

	P-type	N-type
12.	3	5
13.	holes	electrons
14.	electrons	holes

Complete the following chart describing semiconductors.

	a	b	c	d
15.	max	majority	collapsed	0.2 to 0.6
16.	min	minority	expanded	applied V
17.	zero	none	normal	zero

18.

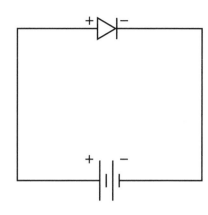

Forward biased

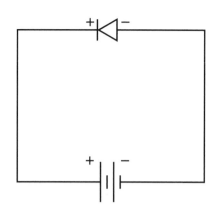

Reverse biased

19.

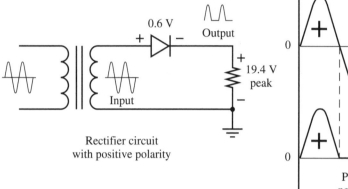

Rectifier circuit
with positive polarity

20.

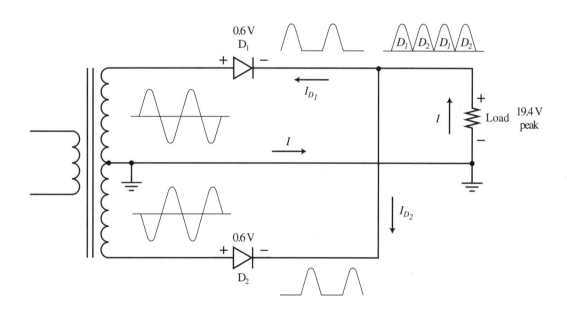

21.

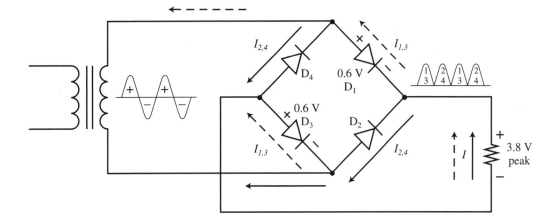

22.

23.   Formula: V_{dc} = peak voltage
       Answer: V_{dc} = 80 V

24.   a.   Zener voltage:
             Formula: $V_Z = V_L$
             Answer: V_Z = 48 V

       b.   Maximum zener current:
             Formula: $I_{Z(max)} = I_L$
             Answer: $I_{Z(max)}$ = 1.5 A

       c.   Zener power rating:
             Formula: $P = I_{Z(max)} \times V_Z$
             Substitution: P = 1.5 A × 48 V
             Answer: P = 72 W

25.   a.   Current through the series resistor:
             Formula: $I_{R_S} = I_{Z(max)}$
             Answer: I_{R_S} = 1.5 A

       b.   Voltage across the series resistor:
             Formula: $V_{R_S} = V_{in} - V_{out}$
             Substitution: V_{R_S} = 60 V – 48 V
             Answer: V_{R_S} = 12 V

       c.   Ohmic value of the series resistor:

             Formula: $R_S = \dfrac{V_{R_S}}{I_{R_S}}$

             Substitution: $R_S = \dfrac{12\ V}{1.5\ A}$

             Answer: R_S = 8 Ω

       d.   Formula: $P_{R_S} = I_{R_S} \times V_{R_S}$
             Substitution: P_{R_S} = 1.5 A × 12 V
             Answer: P_{R_S} = 18 W

Chapter Test
26

Name: _____

Date: _____

Class: _____

Diodes and Power Supplies

Fill in the blanks.

1. The process of changing ac to dc is called _____.

2. A _____ impurity is a doping material with 3 valence electrons.

3. A _____ impurity is a doping material with 5 valence electrons.

4. A _____ bond is formed by the sharing of valence electrons.

5. The _____ region is the area near the junction where the current carriers have been eliminated.

6. _____ bias allows majority current flow.

7. The voltage necessary to overcome the depletion region is the _____ potential.

8. A diode circuit which makes a sine wave look like a square wave is a _____ circuit.

9. A diode circuit that changes the centerline of an ac signal to a dc voltage level is a _____circuit.

10. A zener diode normally operates in the _____ direction.

Select the best answer.

_____ 11. What is meant by the statement, "a diode is a one-way switch"?
   a. A diode conducts current in one direction and blocks the other.
   b. One voltage polarity turns it on, the other polarity turns it off.
   c. The load receives a voltage when the diode is forward biased.
   d. All of the above statements are correct.

Complete the following chart describing semiconductors.

		P-type	N-type
12.	Number of valence electrons in the doping impurity	_____	_____
13.	Majority carriers	_____	_____
14.	Minority carriers	_____	_____

For questions 15 through 17, complete the chart by answering the following questions about a junction diode in the condition listed.

    a. What is the amount of current? (zero, minimum, maximum)

    b. What type of current carriers flow? (none, majority, minority)

    c. In what state is the depletion region? (normal, expanded, collapsed)

    d. What is the voltage drop across the diode? (zero, 0.2 to 0.6, applied voltage)

		a	b	c	d
15.	Forward biased	_____	_____	_____	_____
16.	Reverse biased	_____	_____	_____	_____
17.	No bias	_____	_____	_____	_____

18. Draw the schematic symbol of a diode.

    a. Use + and – symbols to show the correct polarity for forward bias.

    b. Use + and – symbols to show the correct polarity for reverse bias.

19. Draw a basic half-wave rectifier circuit with a 40 volt peak-to-peak sine wave applied. Also draw the output waveform. Use ± to indicate voltage polarities in the circuit.

20. Draw a basic full-wave rectifier circuit with a 40 volt peak-to-peak sine wave applied. Use arrows to indicate current flow.

21. Draw a basic bridge rectifier circuit with a 10 volt peak-to-peak sine wave applied. Use arrows to indicate the direction of the current. Show the expected voltage drops across two of the diodes and the resistor.

22. Draw the output of an unfiltered half-wave rectifier. On this same drawing, show the effect of a simple capacitive power supply filter.

23. Calculate the no load output voltage of a power supply with a capacitive filter on a half-wave rectifier with an input of 80 volts peak.

Formula: _____

Substitution: _____

Answer: _____

For questions 24 and 25, calculate the circuit values of a zener regulator for a 48 volt output at 1.5 amps. The input voltage is 60 volts dc. Note that the given values allow for tolerances.

24. Find the zener diode rating.

a. Zener voltage:

Formula: _____

Substitution: _____

Answer: _____

b. Maximum zener current:

Formula: _____

Substitution: _____

Answer: _____

c. Zener power rating:

Formula: _____

Substitution: _____

Answer: _____

25. Find the values for the series resistor.

   a. Current through the series resistor:

   Formula: _____

   Substitution: _____

   Answer: _____

   b. Voltage across the series resistor:

   Formula: _____

   Substitution: _____

   Answer: _____

c. Ohmic value of the series resistor:

Formula: _____

Substitution: _____

Answer: _____

d. Power dissipated in the series resistor:

Formula: _____

Substitution: _____

Answer: _____

OBJECTIVES

After studying this chapter, students should be able to:
- Define technical terms associated with bipolar junction transistors.
- Test a transistor with an ohmmeter.
- Describe the current relationships in a transistor.
- Use the beta formula to calculate current.
- Describe the voltage relationships in a transistor.
- Define the range of transistor operation.
- Calculate and plot the dc load line on a transistor characteristic curve.
- Calculate the Q point and dc operating parameters for different biasing arrangements.
- Describe how an ac signal passes through a common emitter amplifier.
- Describe the differences between common emitter, common base, and common collector amplifiers.

INSTRUCTIONAL MATERIALS

Text: Pages 791–820
Test Your Knowledge Questions, Pages 818–820
Study Guide: Pages 239–248
Laboratory Manual: Pages 309–322

ANSWERS TO TEXTBOOK

Test Your Knowledge, Pages 818–820
1. N-type.
2. Test each diode in the transistor for proper operation in both the forward and reverse bias state.
3. Emitter.
4. Electrons.
5. Holes.
6. Emitter.
7. Base.
8. $I_C = I_B \times \beta$
9. Emitter current is the sum of the base and collector currents.
10. $I_E = I_C + I_B$
11. $\beta = \dfrac{I_C}{I_B}$
12. $I_E = 10.08$ mA
13. $\beta = 125$
14. 0.6 V for Si.
0.2 V for Ge.
15. From near zero to the full applied voltage.
16. NPN: V_{CC} is + on collector and – on emitter.
PNP: V_{CC} is – on collector and + on emitter.
17. From saturation to cutoff
18. a. $I_{C(sat)} = 4$ mA
$V_{CE} = 20$ V at cutoff
b. Student plot of load line.
c. $I_{C(Q)} = 2$ mA
d. $V_{R_C} = 10$ V
e. $V_{CE(Q)} = 10$ V
f. Student plot of Q point on the load line.
g. $I_{B(Q)} = 20$ μA
19. a. $I_{C(sat)} = 5$ mA
$V_{CE} = 30$ V at cutoff
b. Student plot of load line.
c. $I_{C(Q)} = 2.73$ mA
d. $V_{R_C} = 13.65$ V
e. $V_{R_E} = 2.73$ V
f. $V_{CE(Q)} = 13.62$ V
g. Student plot of Q point on the load line.
20. a. $I_{C(sat)} = 15$ mA
$V_{CE} = 15$ V at cutoff
b. Student plot of load line.
c. $V_{B(Q)} = 2$ V
d. $I_{C(Q)} = 7$ mA
e. $V_{R_C} = 5.6$ V
f. $V_{R_E} = 1.4$ V
g. $V_{CE(Q)} = 8$ V
h. Student plot of Q point on the load line.

21. The ac signal changes the base voltage and base current. This changes collector current, which in turn changes the voltage across the transistor. A larger ac voltage produces a larger collector current, so more voltage is dropped across the collector resistor, leaving less voltage for the load. The resultant sine wave is larger than the input and 180° out of phase.

ANSWERS TO STUDY GUIDE

Pages 239–248

1. b.
2. g.
3. f.
4. c.
5. d.
6. a.
7. h.
8. e.
9. a.
10. a.
11. c.
12. a.
13. b.
14. b.
15. c.
16. a.
17. d.
18. Formula: $I_E = I_B + I_C$
    Substitution: $I_E = 80 \ \mu A + 10 \ mA$
    Answer: $I_E = 10.08 \ mA$

19. Formula: $\beta = \dfrac{I_C}{I_B}$

    Substitution: $\beta = \dfrac{10 \ mA}{80 \ \mu A}$

    Answer: $\beta = 125$

20. a.
21. b.
22. a.
23. a.
24. b.
25. a.

26. a. Calculate I_C at saturation and V_{CE} at cutoff.

    I_C at saturation:

    Formula: $I_{C(sat)} = \dfrac{V_{CC}}{R_C}$

    Substitution: $I_{C(sat)} = \dfrac{20 \ V}{5 \ k\Omega}$

    Answer: $I_{C(sat)} = 4 \ mA$

    V_{CE} at cutoff:

    No formula needed: V_{CE} at cutoff $= V_{CC}$

    Answer: $V_{CE} = 20 \ V$

    b. Student graph of the load line.

    c. I_C at the Q point:
    Note: first find the equivalent resistance of the collector resistor and the transistor.

    Formula: $R_C = \dfrac{R_B}{\beta}$

    Substitution: $R_C = \dfrac{1 \ M\Omega}{100}$

    Answer: $R_C = 10 \ k\Omega$

    Formula: $I_{C(Q)} = \dfrac{V_{CC}}{R_C}$

    Substitution: $I_{C(Q)} = \dfrac{20 \ V}{10 \ k\Omega}$

    Answer: $I_{C(Q)} = 2 \ mA$

    d. V_{R_C} at the Q point:
    Formula: $V_{R_C} = I_{C(Q)} \times R_C$
    Substitution: $V_{R_C} = 2 \ mA \times 5 \ k\Omega$
    Answer: $V_{R_C} = 10 \ V$

    e. V_{CE} at the Q point:
    Formula: $V_{CE(Q)} = V_{CC} - V_{R_C}$
    Substitution: $V_{CE(Q)} = 20 \ V - 10 \ V$
    Answer: $V_{CE(Q)} = 10 \ V$

f. Student plot of the Q point on the load line.

g. I_B at the Q point:

Formula: $I_{B(Q)} = \dfrac{I_C}{\beta}$

Substitution: $I_{B(Q)} = \dfrac{2 \text{ mA}}{100}$

Answer: $I_{B(Q)} = 20 \ \mu\text{A}$

27. a. Calculate I_C at saturation and V_{CE} at cutoff.

I_C at saturation:

Formula: $I_{C(sat)} = \dfrac{V_{CC}}{R_C + R_E}$

Substitution: $I_{C(sat)} = \dfrac{30 \text{ V}}{5 \text{ k}\Omega + 1 \text{ k}\Omega}$

Answer: $I_{C(sat)} = 5 \text{ mA}$

V_{CE} at cutoff:

No formula needed: V_{CE} at cutoff $= V_{CC}$

Answer: $V_{CE} = 30 \text{ V}$

b. Student graph of the load line.

c. I_C at the Q point:
Note: first find the equivalent resistance of the collector resistor and the transistor.

Formula: $R_C = \dfrac{R_B}{\beta}$

Substitution: $R_C = \dfrac{1 \text{ M}\Omega}{100}$

Answer: $R_C = 10 \text{ k}\Omega$

Formula: $I_{C(Q)} = \dfrac{V_{CC}}{R_C + R_E}$

Substitution: $I_{C(Q)} = \dfrac{30 \text{ V}}{10 \text{ k}\Omega + 1 \text{ k}\Omega}$

Answer: $I_{C(Q)} = 2.73 \text{ mA}$

d. V_{R_C} at the Q point:

Formula: $V_{R_C} = I_{C(Q)} \times R_C$

Substitution: $V_{R_C} = 2.73 \text{ mA} \times 5 \text{ k}\Omega$

Answer: $V_{R_C} = 13.65 \text{ V}$

e. V_{R_E} at the Q point:

Formula: $V_{R_E} = I_{C(Q)} \times R_E$

Substitution: $V_{R_E} = 2.73 \text{ mA} \times 1 \text{ k}\Omega$

Answer: $V_{R_E} = 2.73 \text{ V}$

f. V_{CE} at the Q point:

Formula: $V_{CE(Q)} = V_{CC} - (V_{R_C} + V_{R_E})$

Substitution:

$V_{CE(Q)} = 30 \text{ V} - (13.65 \text{ V} + 2.73 \text{ V})$

Answer: $V_{CE(Q)} = 13.62 \text{ V}$

g. Student plot of the Q point on the load line.

28. a. Calculate I_C at saturation and V_{CE} at cutoff.

I_C at saturation:

Formula: $I_{C(sat)} = \dfrac{V_{CC}}{R_C + R_E}$

Substitution: $I_{C(sat)} = \dfrac{15 \text{ V}}{800 \ \Omega + 200 \ \Omega}$

Answer: $I_{C(sat)} = 15 \text{ mA}$

V_{CE} at cutoff:

No formula needed: V_{CE} at cutoff $= V_{CC}$

Answer: $V_{CE} = 15 \text{ V}$

b. Student graph of the load line.

c. Base voltage at the Q point:

Formula: $V_{B(Q)} = V_{CC} \times \dfrac{R_2}{R_1 + R_2}$

Substitution: $V_{B(Q)} = 15 \text{ V} \times \dfrac{1.6 \text{ k}\Omega}{10 \text{ k}\Omega + 1.6 \text{ k}\Omega}$

Answer: $V_{B(Q)} = 2 \text{ V}$

d. I_C at the Q point:

First find V_{R_E}:

Formula: $V_{R_E} = V_B - V_{BE}$

Substitution: $V_{R_E} = 2\text{ V} - 0.6\text{ V}$

Answer: $V_{R_E} = 1.4\text{ V}$

Collector current:

Formula: $I_{C(Q)} = \dfrac{V_{R_E}}{R_E}$

Substitution: $I_{C(Q)} = \dfrac{1.4\text{ V}}{200\ \Omega}$

Answer: $I_{C(Q)} = 7\text{ mA}$

e. V_{R_C} at the Q point:

Formula: $V_{R_C} = I_{C(Q)} \times R_C$

Substitution: $V_{R_C} = 7\text{ mA} \times 800\ \Omega$

Answer: $V_{R_C} = 5.6\text{ V}$

f. V_{R_E} at the Q point:

Formula: $V_{R_E} = I_{C(Q)} \times R_E$

Substitution: $V_{R_E} = 7\text{ mA} \times 200\ \Omega$

Answer: $V_{R_E} = 1.4\text{ V}$

g. V_{CE} at the Q point:

Formula: $V_{CE(Q)} = V_{CC} - (V_{R_C} + V_{R_E})$

Substitution: $V_{CE(Q)} = 15\text{ V} - (5.6\text{ V} + 1.4\text{ V})$

Answer: $V_{CE(Q)} = 8\text{ V}$

h. Student plot of the Q point on the load line.

29. Refer to textbook figure 27-20.

30.

	CE	CB	CC
Input sine wave	B	E	B
Output sine wave	C	C	E
Phase shift	180°	0°	0°
Amplification	good	good	< 1

ANSWERS TO CHAPTER TEST IN THE INSTRUCTOR'S MANUAL

Pages 351–358

In the chart, "X" is the correct answer. The remainder of chart should remain blank.

	Base	Emitter	Collector
1.	X		
2.			X
3.		X	
4.		X	
5.	X		

In the chart, "X" is the correct answer. The remainder of chart should remain blank.

	active region	cutoff	saturation	Q point
6.	X			
7.				X
8.			X	
9.		X		

10. a.

11. b.

12. a.

13. d.

14. Formula: $I_E = I_B + I_C$

Substitution: $I_E = 150\ \mu\text{A} + 30\text{ mA}$

Answer: $I_E = 30.15\text{ mA}$

15. Formula: $\beta = \dfrac{I_C}{I_B}$

Substitution: $\beta = \dfrac{30\text{ mA}}{150\ \mu\text{A}}$

Answer: $\beta = 200$

16. a. 0.6 V

  b. 0.2 V

17. 0 V, V_{CC}

18. a.

19. b.

20. a.

21. a. I_C at saturation:

   Formula: $I_{C(sat)} = \dfrac{V_{CC}}{R_L}$

   Substitution: $I_{C(sat)} = \dfrac{30\text{ V}}{4\text{ k}\Omega}$

   Answer: $I_{C(sat)} = 7.5\text{ mA}$

   b. V_{CE} at cutoff:
   Formula: $V_{CE} = V_{CC}$
   Answer: $V_{CE} = 30\text{ V}$

   c.

   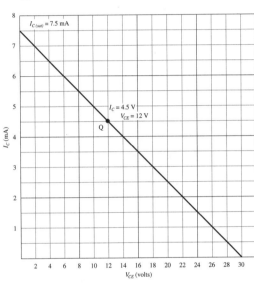

22. a. Calculate the collector resistance at the Q point, including the transistor.

   Formula: $R_{C(Q)} = \dfrac{R_B}{\beta}$

   Substitution: $R_{C(Q)} = \dfrac{1\text{ M}\Omega}{150}$

   Answer: $R_{C(Q)} = 6.67\text{ k}\Omega$

   b. Calculate the collector current at the Q point.

   Formula: $I_{C(Q)} = \dfrac{V_{CC}}{R_{C(Q)}}$

   Substitution: $I_{C(Q)} = \dfrac{30\text{ V}}{6.67\text{ k}\Omega}$

   Answer: $I_{C(Q)} = 4.5\text{ mA}$

   c. Calculate the voltage drop across the collector resistor.

   Formula: $V_{R_L} = I_{C(Q)} \times R_L$
   Substitution: $V_{R_L} = 4.5\text{ mA} \times 4\text{ k}\Omega$
   Answer: $V_{R_L} = 18\text{ V}$

   d. Calculate V_{CE} at the Q point.
   Formula: $V_{CE(Q)} = V_{CC} - V_{R_L}$
   Substitution: $V_{CE(Q)} = 30\text{ V} - 18\text{ V}$
   Answer: $V_{CE(Q)} = 12\text{ V}$

23. See graph in answer to question 21, part c.

24. Formula: $I_{B(Q)} = \dfrac{I_{C(Q)}}{\beta}$

   Substitution: $I_{B(Q)} = \dfrac{4.5\text{ mA}}{150}$

   Answer: $I_{B(Q)} = 30\text{ μA}$

25. Complete the chart.

	CE	CB	CC
Input sine wave	B	E	B
Output sine wave	C	C	E
Phase shift	180°	0°	0°
Voltage amplification	good	good	<1

Chapter Test

27

Name: _____

Date: _____

Class: _____

Bipolar Junction Transistors

In the following chart, place an "X" for the element that answers the statement for a bipolar junction transistor.

		Base	Emitter	Collector
1.	Controls the current.	_____	_____	_____
2.	Output of the transistor.	_____	_____	_____
3.	Supplies majority carriers.	_____	_____	_____
4.	Contains largest current.	_____	_____	_____
5.	Contains smallest current.	_____	_____	_____

In the following chart, place an "X" for the operating characteristic that matches the statement for a bipolar junction transistor.

		active region	cutoff	saturation	Q point
6.	Acts like a variable resistor.	_____	_____	_____	_____
7.	The actual operating juncture.	_____	_____	_____	_____
8.	Acts like a closed switch.	_____	_____	_____	_____
9.	Acts like an open switch.	_____	_____	_____	_____

Select the best answer.

_____ 10. In an NPN transistor the majority current is:
   a. electrons.
   b. holes.

_____ 11. In a PNP transistor the majority current is:
   a. electrons.
   b. holes.

_____ 12. What is the formula for emitter current, in terms of base and collector currents?

    a. $I_E = I_C + I_B$

    b. $I_E = I_C - I_B$

    c. $I_E = I_C \times I_B$

    d. $I_E = I_C / I_B$

13. What is the beta formula?

    a. $\beta = I_C + I_B$

    b. $\beta = I_C - I_B$

    c. $\beta = I_C \times I_B$

    d. $\beta = I_C / I_B$

With each problem, write the formula, substitution, and answer.

14. Calculate emitter current with a base current of 150 μA and collector current of 30 mA.

Formula: _____

Substitution: _____

Answer: _____

15. Calculate the beta of the transistor in question 14.

Formula: _____

Substitution: _____

Answer: _____

Fill in the blanks.

16. What is the approximate voltage V_{BE} for:

    a. a silicon transistor? _____

    b. a germanium transistor? _____

17. The range of operating voltages for V_{CE} is from _____ to _____ (approximately).

Select the best answer.

_____ 18. The polarity for V_{CC} with an NPN transistor is:
    a. + on collector, − on emitter.
    b. − on collector, + on emitter.
    c. − on base, + on emitter.
    d. + on base, − on collector.

_____ 19. The polarity for V_{CC} with a PNP transistor is:
    a. + on collector, − on emitter.
    b. − on collector, + on emitter.
    c. + on base, − on emitter.
    d. − on base, + on collector.

_____ 20. A transistor operates in a range from:
    a. cutoff to saturation.
    b. cutoff to zero collector current.
    c. saturation to maximum collector current.
    d. saturation to maximum base current.

For questions 21 through 24, solve for the requested values using the following circuit.

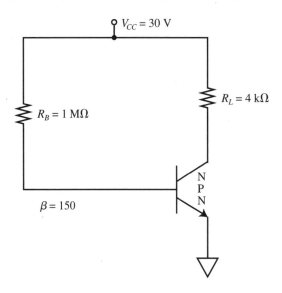

21. Calculate the dc load line.

   a. I_C at saturation:

   Formula: _____

   Substitution: _____

   Answer: _____

b. V_{CE} at cutoff:

Formula: _____

Substitution: _____

Answer: _____

c. Plot the load line on graph paper.

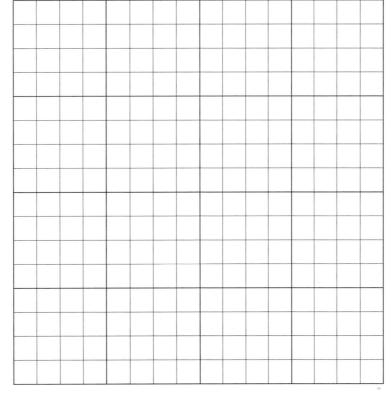

V_{CE} (volts)

22. Calculate the Q point.

   a. Calculate the collector resistance at the Q point, including the transistor.

   Formula: _____

   Substitution: _____

   Answer: _____

   b. Calculate the collector current at the Q point.

   Formula: _____

   Substitution: _____

   Answer: _____

c. Calculate the voltage drop across the collector resistor.

Formula: _____

Substitution: _____

Answer: _____

d. Calculate V_{CE} at the Q point.

Formula: _____

Substitution: _____

Answer: _____

23. Plot the Q point. (Use the load line graph created in question 21.)

24. Calculate base current at the Q point.

Formula: _____

Substitution: _____

Answer: _____

25. Complete the chart comparing common emitter, common base, and common collector amplifiers.

	CE	*CB*	*CC*
Input sine wave (B, C, or E)	_____	_____	_____
Output sine wave (B, C, or E)	_____	_____	_____
Phase shift (0° or 180°)	_____	_____	_____
Voltage amplification (good or <1)	_____	_____	_____

Other Semiconductor Devices

OBJECTIVES

After studying this chapter, students should be able to:

- Define technical terms related to semiconductor devices.
- Describe the operation of field-effect transistors.
- Describe the operation of basic operational amplifiers (op-amps) and calculate output gain.
- Describe the operation of the 555 timer as an integrated circuit and calculate the output waveform frequency and duty cycle.
- Describe the operation of the unijunction transistor (UJT).
- Describe the operation of the silicon controlled rectifier (SCR).
- State typical applications for the DIAC, TRIAC, LED, and light-sensitive transistor.

INSTRUCTIONAL MATERIALS

Text: Pages 821–842
Test Your Knowledge Questions, Pages 841–842
Study Guide: Pages 249–256
Laboratory Manual: Pages 323–340

ANSWERS TO TEXTBOOK

Test Your Knowledge, Pages 841–842

1. Junction field-effect transistors (JFET), insulated-gate field-effect transistors (IGFET).
2. Refer to figure 28-1. The gate opens and closes the channel allowing current to travel between the source and drain.
3. Refer to figure 28-2. When there is no gate voltage, current flows through the channel with no opposition. As the gate voltage increases, majority carriers from the gate are forced into the channel. This produces an area that opposes current in the channel.
4. Small gate current. Maximum drain current.
5. Large gate current. Zero drain current.
6. Medium gate current. Medium drain current.
7. Metal.
8. Semiconductor.
9. A place for the displaced majority carriers to go.
10. A supply for majority carriers.
11. Increased gate voltage depletes the channel and reduces current.
12. Increased gate voltage enhances the channel and increases current.
13. Refer to figure 28-6.
14. Refer to figure 28-6.
15. Refer to figure 28-8.
16. $A_V = 5$
17. Refer to figure 28-9.
18. $A_V = 6$
19. Reduces gain and stabilizes the amplifier. R_F.
20. Refer to figure 28-11.
21. Rectangular waveform or square wave.
22. $f = 265$ Hz
23. $C = 5830$ pF
24. Duty cycle = 60%
25. Timing circuit.
26. Refer to figure 28-13.
27. Refer to figure 28-14. Both outputs have the shape of an RC time constant circuit.
28. Refer to figure 28-15.
29. The gate activates the SCR. The holding current keeps it activated.
30. Refer to figure 28-16.
31. To control ac voltages. SCR.
32. A voltage reference. Zener diode.
33. Light-emitting diode. A small indicator lamp.
34. A transistor where light activates the base. A light-activated switch.

ANSWERS TO STUDY GUIDE

Pages 249–256

1. k.
2. d.
3. f.
4. b.
5. j.
6. a.
7. i.
8. l.
9. c.
10. e.
11. g.
12. h.
13. m.
14. Refer to textbook figure 28-1.
15. Refer to textbook figure 28-3.
16. Refer to textbook figure 28-2.
17. Refer to textbook figure 28-3.
18. b.
19. a.
20. a.
21. b.
22. c.
23. c.
24. d.
25. b.
26. a.
27. b.
28. b.
29. a.
30. Refer to textbook figure 28-6.
31. Refer to textbook figure 28-6.
32. Refer to textbook figure 28-8.

33. Formula: $A_V = \dfrac{R_F}{R_I}$

   Substitution: $A_V = \dfrac{20\ \text{k}\Omega}{5\ \text{k}\Omega}$

   Answer: $A_V = 4$

34. Refer to textbook figure 28-9.

35. Formula: $A_V = 1 + \dfrac{R_F}{R_I}$

   Substitution: $A_V = 1 + \dfrac{25\ \text{k}\Omega}{5\ \text{k}\Omega}$

   Answer: $A_V = 6$

36. b.
37. Refer to textbook figure 28-11.
38. a.

39. Formula: $f = \dfrac{1.44}{(R_A + 2R_B) \times C_I}$

   Substitution:

   $$f = \dfrac{1.44}{(1\ \text{k}\Omega + 2 \times 3.3\ \text{k}\Omega) \times 0.001\ \mu\text{F}}$$

   Answer: $f = 189$ kHz

40. Formula: $C_I = \dfrac{1.44}{(R_A + 2R_B) \times f}$

   Substitution:

   $$C_I = \dfrac{1.44}{(1\ \text{k}\Omega + 2 \times 3.3\ \text{k}\Omega) \times 10\ \text{kHz}}$$

   Answer: $C_I = 0.018\ \mu\text{F}$

41. Formula: Duty cycle $= \dfrac{R_A + R_B}{R_A + 2R_B} \times 100\%$

   Substitution:

   Duty cycle $= \dfrac{1\ \text{k}\Omega + 3.3\ \text{k}\Omega}{1\ \text{k}\Omega + 2(3.3\ \text{k}\Omega)} \times 100\%$

   Answer: Duty cycle $= 56.6\%$

42. c.
43. Refer to textbook figure 28-13.
44. Refer to textbook figure 28-14.
45. Refer to textbook figure 28-15.
46. a.
47. b.
48. Refer to textbook figure 28-16.
49. c.
50. c.
51. b.
52. a.

ANSWERS TO CHAPTER TEST IN THE INSTRUCTOR'S MANUAL

Pages 363–366

1.

BJT	JFET
Emitter	Source
Collector	Drain
Base	Gate

2. a. Saturation.
    b. Active.
    c. Pinch-off.
3. substrate
4. depletion
5. enhancement
6. inverting
7. noninverting
8. virtual ground
9. *Complete the chart.*

	Saturation	*Pinch-off*	*Active*
Gate voltage	none	large	varies
Drain current	large	none	varies

10. b.
11. a.
12.

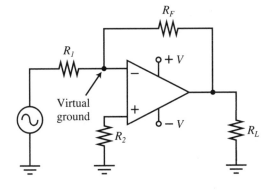

13. Formula: $A_V = \dfrac{R_F}{R_I}$

    Substitution: $A_V = \dfrac{50 \text{ k}\Omega}{2.5 \text{ k}\Omega}$

    Answer: $A_V = 20$

14.

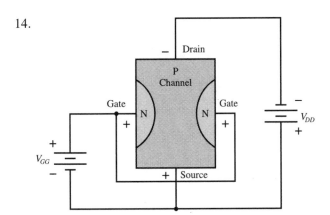

15.

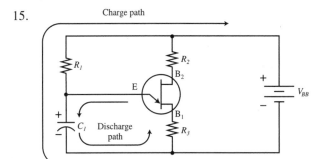

16.

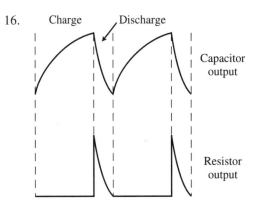

17.

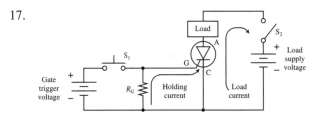

18. a.
19. b.
20. c.

Chapter Test

28

Name: _____

Date: _____

Class: _____

Other Semiconductor Devices

1. Place the names of the three elements of a JFET next to the comparable elements from a BJT.

BJT	*JFET*
Emitter	_____
Collector	_____
Base	_____

2. Identify the range of operation for a JFET, based on the following descriptions.

   a. No gate voltage is applied and there is maximum current: _____

   b. Amplifies a sine wave with minimum distortion: _____

   c. No current flows: _____

Fill in the blanks.

3. The _____ is a massive block of semiconductor material made of the opposite type

   from the channel of a MOSFET.

4. The _____ mode of operation is when the gate voltage has the same polarity as the

   channel forcing the current carriers out of the channel of a MOSFET.

5. The _____ mode of operation is when the gate voltage has the same polarity as the sub-

   strate, supplying current carriers to form a channel in a MOSFET.

6. When the output signal of an op-amp has a phase shift of 180° and the gain is high, the input signal is applied to the

   _____ input.

7. When the output signal of an op-amp has a phase shift of 0° and the gain is slightly higher than one, the input signal

   is applied to the _____ input.

8. The point in an op-amp circuit that appears as zero volts to the ac signal is referred to as: _____

_____.

9. What are the relative amounts of gate voltage and drain current for the operating regions of a JFET? (Answer with: none, varies, or large.)

	Saturation	*Pinch-off*	*Active*
Gate voltage	_____	_____	_____
Drain current	_____	_____	_____

Select the best answer.

_____ 10. When voltage is applied to the gate of a depletion-mode MOSFET, the effect on the channel is to:
   a. increase current from source to drain.
   b. decrease current from source to drain.
   c. increase gate current.
   d. decrease gate current.

_____ 11. When voltage is applied to the gate of an enhancement-mode MOSFET, the effect on the channel is to:
   a. increase current from source to drain.
   b. decrease current from source to drain.
   c. increase gate current.
   d. decrease gate current.

12. Draw the schematic diagram of a basic op-amp circuit with an inverting input. Label the resistors as: R_F, R_1, R_2, R_L. Label the virtual ground point.

13. With the circuit of question 12, calculate the output gain if $R_F = 50$ kΩ and $R_I = 2.5$ kΩ.

Formula: _____

Substitution: _____

Answer: _____

14. Draw the block diagram of a P channel JFET. Include biasing voltages.

15. Draw the schematic diagram of a UJT relaxation oscillator circuit. Label R_1, R_2, R_3, C_1, E, B$_1$, B$_2$, and V_{BB}. Draw arrows to indicate the charge and discharge paths.

16. Draw the expected output waveforms of the circuit in question 15. Draw the outputs for both the output resistor and capacitor.

17. Draw the schematic diagram of an SCR used as a relay in a dc circuit. Label the gate, cathode, and anode. Draw an arrow to indicate the path for load current.

Select the best answer.

_____ 18. The function of the gate in the SCR circuit of question 17 is to:
  a. trigger the anode/cathode circuit.
  b. supply current in the gate circuit.
  c. turn the SCR off when current flows.
  d. allow the SCR to pass a sine wave.

_____ 19. The purpose of the holding current is to:
  a. supply current to the load.
  b. keep the gate on.
  c. keep the gate off until a signal is applied.

_____ 20. The typical application of a TRIAC is:
  a. to control the on/off time of dc circuits.
  b. as a voltage reference.
  c. to control the on/off time of sine circuits.
  d. as a bridge rectifier.

Introduction to Digital Electronics

OBJECTIVES

After studying this chapter, students should be able to:
- Contrast digital circuits with analog circuits.
- Define the binary numbering system for use in digital circuits.
- Convert decimal numbers to binary and binary numbers to decimal.
- Analyze basic binary logic gates such as AND, OR, NOT, NAND, and NOR.

INSTRUCTIONAL MATERIALS

Text: Pages 843–856

Test Your Knowledge Questions, Pages 855–856

Study Guide: Pages 257–262

Laboratory Manual: Pages 341–350

ANSWERS TO TEXTBOOK

Test Your Knowledge, Pages 855–856

1. Analog circuits have a continuous range of voltages. Digital circuits operate using only discrete steps. They often have only two voltage states, on and off.

2. $10^0 = 1$
   $10^1 = 10$
   $10^2 = 100$
   $10^3 = 1,000$
   $10^4 = 10,000$
   $10^5 = 100,000$
   $10^6 = 1,000,000$

3. a. $4,302,916 =$
   $4 \times 1,000,000 = 4,000,000$
   $3 \times 100,000 = 300,000$
   $0 \times 10,000 = 0$
   $2 \times 1,000 = 2,000$
   $9 \times 100 = 900$
   $1 \times 10 = 10$
   $6 \times 1 = 6$

   b. $57,008 =$
   $5 \times 10,000 = 50,000$
   $7 \times 1,000 = 7,000$
   $0 \times 100 = 0$
   $0 \times 10 = 0$
   $8 \times 1 = 8$

   c. $256 =$
   $2 \times 100 = 200$
   $5 \times 10 = 50$
   $6 \times 1 = 6$

   d. $7,029 =$
   $7 \times 1,000 = 7,000$
   $0 \times 100 = 0$
   $2 \times 10 = 20$
   $9 \times 1 = 9$

   e. $2,682,530 =$
   $2 \times 1,000,000 = 2,000,000$
   $6 \times 100,000 = 600,000$
   $8 \times 10,000 = 80,000$
   $2 \times 1,000 = 2,000$
   $5 \times 100 = 500$
   $3 \times 10 = 30$
   $0 \times 1 = 0$

4. $2^0 = 1$
   $2^1 = 2$
   $2^2 = 4$
   $2^3 = 8$
   $2^4 = 16$
   $2^5 = 32$
   $2^6 = 64$
   $2^7 = 128$
   $2^8 = 256$

5. Refer to figure 29-3.
6. a. $101110_2 = 46_{10}$
   $1 \times 32 = 32$
   $0 \times 16 = 0$
   $1 \times 8 = 8$
   $1 \times 4 = 4$
   $1 \times 2 = 2$
   $0 \times 1 = 0$
   b. $11101_2 = 29_{10}$
   $1 \times 16 = 16$
   $1 \times 8 = 8$
   $1 \times 4 = 4$
   $0 \times 2 = 0$
   $1 \times 1 = 1$
   c. $1100110_2 = 102_{10}$
   $1 \times 64 = 64$
   $1 \times 32 = 32$
   $0 \times 16 = 0$
   $0 \times 8 = 0$
   $1 \times 4 = 4$
   $1 \times 2 = 2$
   $0 \times 1 = 0$
   d. $00111001_2 = 57_{10}$
   $0 \times 128 = 0$
   $0 \times 64 = 0$
   $1 \times 32 = 32$
   $1 \times 16 = 16$
   $1 \times 8 = 8$
   $0 \times 4 = 0$
   $0 \times 2 = 0$
   $1 \times 1 = 1$
   e. $10101011_2 = 171_{10}$
   $1 \times 128 = 128$
   $0 \times 64 = 0$
   $1 \times 32 = 32$
   $0 \times 16 = 0$
   $1 \times 8 = 8$
   $0 \times 4 = 0$
   $1 \times 2 = 2$
   $1 \times 1 = 1$
7. $29_{10} = 11101_2$
   $256_{10} = 100000000_2$
   $150_{10} = 10010110_2$
   $100_{10} = 1100100_2$
   $35_{10} = 100011_2$
8. $101_2 = 5_{10}$
   $110011_2 = 51_{10}$
   $101100_2 = 44_{10}$
   $1110111_2 = 119_{10}$
   $10101010_2 = 170_{10}$

9. AND—Refer to figure 29-7.
   OR—Refer to figure 29-10.
   NOT—Refer to figure 29-13.
   NAND—Refer to figure 29-16.
   NOR—Refer to figure 29-19.
10. AND—Refer to figure 29-8.
    OR—Refer to figure 29-11.
    NOT—Refer to figure 29-14.
    NAND—Refer to figure 29-17.
    NOR—Refer to figure 29-20.

ANSWERS TO STUDY GUIDE

Pages 257–262
1. a.
2. e.
3. d.
4. b.
5. c.
6. $10^0 = 1$
   $10^1 = 10$
   $10^2 = 100$
   $10^3 = 1,000$
   $10^4 = 10,000$
   $10^5 = 100,000$
   $10^6 = 1,000,000$
7. $87,090 =$
   $8 \times 10^4 = 80,000$
   $7 \times 10^3 = 7,000$
   $0 \times 10^2 = 0$
   $9 \times 10^1 = 90$
   $0 \times 10^0 = 0$
8. $350,012 =$
   $3 \times 10^5 = 300,000$
   $5 \times 10^4 = 50,000$
   $0 \times 10^3 = 0$
   $0 \times 10^2 = 0$
   $1 \times 10^1 = 10$
   $2 \times 10^0 = 2$
9. $256 =$
   $2 \times 10^2 = 200$
   $5 \times 10^1 = 50$
   $6 \times 10^0 = 6$

10. $7{,}050{,}812 =$
      $7 \times 10^6 = 7{,}000{,}000$
      $0 \times 10^5 = \quad 0$
      $5 \times 10^4 = \quad 50{,}000$
      $0 \times 10^3 = \quad 0$
      $8 \times 10^2 = \quad 800$
      $1 \times 10^1 = \quad 10$
      $2 \times 10^0 = \quad 2$

11. $6{,}014 =$
      $6 \times 10^3 = \quad 6{,}000$
      $0 \times 10^2 = \quad 0$
      $1 \times 10^1 = \quad 10$
      $4 \times 10^0 = \quad 4$

12. $2^0 = 1$
      $2^1 = 2$
      $2^2 = 4$
      $2^3 = 8$
      $2^4 = 16$
      $2^5 = 32$
      $2^6 = 64$
      $2^7 = 128$
      $2^8 = 256$

13. $101110_2 = 46_{10}$
      $1 \times 32 = 32$
      $0 \times 16 = 0$
      $1 \times 8 = 8$
      $1 \times 4 = 4$
      $1 \times 2 = 2$
      $0 \times 1 = 0$

14. $11101_2 = 29_{10}$
      $1 \times 16 = 16$
      $1 \times 8 = 8$
      $1 \times 4 = 4$
      $0 \times 2 = 0$
      $1 \times 1 = 1$

15. $1100110_2 = 102_{10}$
      $1 \times 64 = 64$
      $1 \times 32 = 32$
      $0 \times 16 = 0$
      $0 \times 8 = 0$
      $1 \times 4 = 4$
      $1 \times 2 = 2$
      $0 \times 1 = 0$

16. $00111001_2 = 57_{10}$
      $0 \times 128 = 0$
      $0 \times 64 = 0$
      $1 \times 32 = 32$
      $1 \times 16 = 16$
      $1 \times 8 = 8$
      $0 \times 4 = 0$
      $0 \times 2 = 0$
      $1 \times 1 = 1$

17. $10101011_2 = 171_{10}$
      $1 \times 128 = 128$
      $0 \times 64 = 0$
      $1 \times 32 = 32$
      $0 \times 16 = 0$
      $1 \times 8 = 8$
      $0 \times 4 = 0$
      $1 \times 2 = 2$
      $1 \times 1 = 1$

18. $29_{10} = 11101_2$
      $150_{10} = 10010110_2$
      $35_{10} = 100011_2$
      $256_{10} = 100000000_2$
      $100_{10} = 1100100_2$

19. $101_2 = 5_{10}$
      $110011_2 = 51_{10}$
      $101100_2 = 44_{10}$
      $1110111_2 = 119_{10}$
      $10101010_2 = 170_{10}$

20.   AND—Refer to figures 29-7 and 29-8.
      OR—Refer to figures 29-10 and 29-11.
      NOT—Refer to figures 29-13 and 29-14.
      NAND—Refer to figures 29-16 and 29-17.
      NOR—Refer to figures 29-19 and 29-20.

ANSWERS TO CHAPTER TEST IN THE INSTRUCTOR'S MANUAL

Pages 373–376

1.   Complete the chart.

10^6	10^5	10^4	10^3	10^2	10^1	10^0
1,000,000	100,000	10,000	1,000	100	10	1

Complete the chart for questions 2 and 3.

	2^5	2^4	2^3	2^2	2^1	2^0
2.	100000	10000	1000	100	10	1
3.	32	16	8	4	2	1

4. $59{,}103 = 5 \times 10^4 = 50{,}000$
      $9 \times 10^3 = \quad 9000$
      $1 \times 10^2 = \quad 100$
      $0 \times 10^1 = \quad 0$
      $3 \times 10^0 = \quad 3$

5. $125,210 = 1 \times 10^5 = 100,000$
     $\phantom{125,210 = {}}2 \times 10^4 = 20,000$
     $\phantom{125,210 = {}}5 \times 10^3 = 5000$
     $\phantom{125,210 = {}}2 \times 10^2 = 200$
     $\phantom{125,210 = {}}1 \times 10^1 = 10$
     $\phantom{125,210 = {}}0 \times 10^0 = 0$

6. $385 = 3 \times 10^2 = 300$
     $\phantom{385 = {}}8 \times 10^1 = 80$
     $\phantom{385 = {}}5 \times 10^0 = 5$

7. $3,450,619 = 2 \times 10^6 = 3,000,000$
     $\phantom{3,450,619 = {}}4 \times 10^5 = 400,000$
     $\phantom{3,450,619 = {}}5 \times 10^4 = 50,000$
     $\phantom{3,450,619 = {}}0 \times 10^3 = 0$
     $\phantom{3,450,619 = {}}6 \times 10^2 = 600$
     $\phantom{3,450,619 = {}}1 \times 10^1 = 10$
     $\phantom{3,450,619 = {}}9 \times 10^0 = 9$

8. $7401 = 7 \times 10^3 = 7000$
     $\phantom{7401 = {}}4 \times 10^2 = 400$
     $\phantom{7401 = {}}0 \times 10^1 = 0$
     $\phantom{7401 = {}}1 \times 10^0 = 1$

9. $110011 = 1 \times 2^5 = 32$
     $\phantom{110011 = {}}1 \times 2^4 = 16$
     $\phantom{110011 = {}}0 \times 2^3 = 0$
     $\phantom{110011 = {}}0 \times 2^2 = 0$
     $\phantom{110011 = {}}1 \times 2^1 = 2$
     $\phantom{110011 = {}}1 \times 2^0 = 1$
     110011 (binary) $= 51$ (decimal)

10. $10110 = 1 \times 2^4 = 16$
     $\phantom{10110 = {}}0 \times 2^3 = 0$
     $\phantom{10110 = {}}1 \times 2^2 = 4$
     $\phantom{10110 = {}}1 \times 2^1 = 2$
     $\phantom{10110 = {}}0 \times 2^0 = 0$
     10110 (binary) $= 22$ (decimal)

11. $1011011 = 1 \times 2^6 = 64$
     $\phantom{1011011 = {}}0 \times 2^5 = 0$
     $\phantom{1011011 = {}}1 \times 2^4 = 16$
     $\phantom{1011011 = {}}1 \times 2^3 = 8$
     $\phantom{1011011 = {}}0 \times 2^2 = 0$
     $\phantom{1011011 = {}}1 \times 2^1 = 2$
     $\phantom{1011011 = {}}1 \times 2^0 = 1$
     1011011 (binary) $= 91$ (decimal)

12. $011101 = 0 \times 2^5 = 0$
     $\phantom{011101 = {}}1 \times 2^4 = 16$
     $\phantom{011101 = {}}1 \times 2^3 = 8$
     $\phantom{011101 = {}}1 \times 2^2 = 4$
     $\phantom{011101 = {}}0 \times 2^1 = 0$
     $\phantom{011101 = {}}1 \times 2^0 = 1$
     011101 (binary) $= 29$ (decimal)

13. $0101110 = 0 \times 2^6 = 0$
     $\phantom{0101110 = {}}1 \times 2^5 = 32$
     $\phantom{0101110 = {}}0 \times 2^4 = 0$
     $\phantom{0101110 = {}}1 \times 2^3 = 8$
     $\phantom{0101110 = {}}1 \times 2^2 = 4$
     $\phantom{0101110 = {}}1 \times 2^1 = 2$
     $\phantom{0101110 = {}}0 \times 2^0 = 0$
     1011011 (binary) $= 46$ (decimal)

14. $25 = 11001$
     $178 = 10110010$
     $46 = 101110$
     $262 = 100000110$
     $110 = 1101110$

15. $1111 = 15$
     $101011 = 43$
     $1101100 = 108$
     $1000001 = 65$
     $11111101 = 253$

16. AND

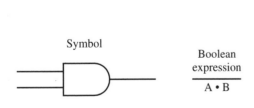

	Inputs		Output
	A	B	A • B
	0	0	0
	0	1	0
	1	0	0
	1	1	1

Symbol

Boolean
expression
A • B

Written
expression

Both A and B
must be present
to produce an output.

17. OR

Symbol

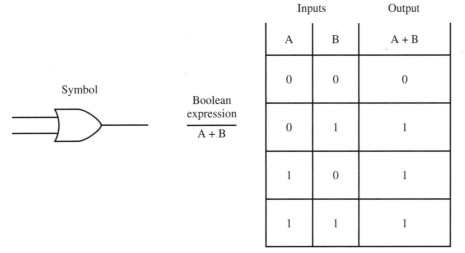

Boolean
expression

$\overline{A + B}$

Inputs		Output
A	B	A + B
0	0	0
0	1	1
1	0	1
1	1	1

Written
expression

Either A or B
or both
will produce
an output.

18. NOT

Symbol

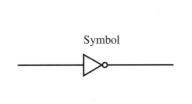

Boolean
expression

$\overline{A}$

Input	Output
A	$\overline{A}$
0	1
1	0

Written
expression

The output is
NOT equal to
the input.

19. NAND

Symbol

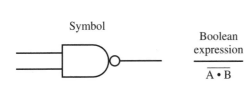

Boolean
expression

$\overline{A \cdot B}$

Inputs		Output
A	B	$\overline{A \cdot B}$
0	0	1
0	1	1
1	0	1
1	1	0

Written
expression

With both A and B
present the
output is low.

20.   NOR

Symbol

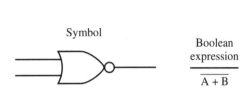

Boolean
expression

$$\overline{A + B}$$

Inputs		Output
A	B	$\overline{A + B}$
0	0	1
0	1	0
1	0	0
1	1	0

Written
expression

With either
A or B the
output is low.

Chapter Test

29

Name: _____

Date: _____

Class: _____

Introduction to Digital Electronics

1. Write a decimal place value chart up to 10^6.

10^6	10^5	10^4	10^3	10^2	10^1	10^0

For questions 2 and 3, use the chart that follows.

2^5	2^4	2^3	2^2	2^1	2^0

2. Fill in the binary number for each binary place value in the chart.

3. Fill in the decimal value equivalent for each binary place value.

For questions 4 through 8, write each digit in the following decimal numbers multiplied by its place value.

Example: $4{,}302{,}916 = 4 \times 10^6 = 4{,}000{,}000$

$$3 \times 10^5 = \quad 300{,}000$$
$$0 \times 10^4 = \quad\quad\quad 0$$
$$2 \times 10^3 = \quad\quad 2000$$
$$9 \times 10^2 = \quad\quad\; 900$$
$$1 \times 10^1 = \quad\quad\quad 10$$
$$6 \times 10^0 = \quad\quad\quad\; 6$$

4. $59{,}103 = 5 \times$ _____ = _____

   $9 \times$ _____ = _____

   $1 \times$ _____ = _____

   $0 \times$ _____ = _____

   $3 \times$ _____ = _____

5. 125,210 = 1 × _____ = _____

2 × _____ = _____

5 × _____ = _____

2 × _____ = _____

1 × _____ = _____

0 × _____ = _____

6. 385 = 3 × _____ = _____

8 × _____ = _____

5 × _____ = _____

7. 3,450,619 = 3 × _____ = _____

4 × _____ = _____

5 × _____ = _____

0 × _____ = _____

6 × _____ = _____

1 × _____ = _____

9 × _____ = _____

8. 7401 = 7 × _____ = _____

4 × _____ = _____

0 × _____ = _____

1 × _____ = _____

For questions 9 through 13, write each digit in the following binary numbers multiplied by its place value. Also, find the decimal equivalent of each.

Example: $101 = 1 \times 2^2 = 4$
$0 \times 2^1 = 0$
$1 \times 2^0 = 1$
101 (binary) = 5 (decimal)

9. 110011 = 1 × _____ = _____

1 × _____ = _____

0 × _____ = _____

0 × _____ = _____

1 × _____ = _____

1 × _____ = _____

110011 (binary) = _____ (decimal)

10. 10110 = 1 × _____ = _____

0 × _____ = _____

1 × _____ = _____

1 × _____ = _____

0 × _____ = _____

10110 (binary) = _____ (decimal)

11. 1011011 = 1 × _____ = _____

0 × _____ = _____

1 × _____ = _____

1 × _____ = _____

0 × _____ = _____

1 × _____ = _____

1 × _____ = _____

1011011 (binary) = _____ (decimal)

12. 011101 = 0 × _____ = _____

1 × _____ = _____

1 × _____ = _____

1 × _____ = _____

0 × _____ = _____

1 × _____ = _____

011101 (binary) = _____ (decimal)

13. 0101110 = 0 × _____ = _____

1 × _____ = _____

0 × _____ = _____

1 × _____ = _____

1 × _____ = _____

1 × _____ = _____

0 × _____ = _____

0101110 (binary) = _____ (decimal)

14. Convert the following decimal numbers to binary.

25 = _____

178 = _____

46 = _____

262 = _____

110 = _____

15. Convert the following binary numbers to decimal.

1111 = _____

101011 = _____

1101100 = _____

1000001 = _____

11111101 = _____

For questions 16 through 20, complete the chart for each binary logic gate.

16. AND

Symbol	Boolean	Truth table			Written expression
		A	B	output	
		0	0	_____	
		0	1	_____	
		1	0	_____	
		1	1	_____	

17. OR

Symbol	Boolean	Truth table			Written expression
		A	B	output	
		0	0	_____	
		0	1	_____	
		1	0	_____	
		1	1	_____	

18. NOT

Symbol	Boolean	Truth table		Written expression
		A	output	
		0	_____	
		1	_____	

19. NAND

Symbol	Boolean	Truth table			Written expression
		A	B	output	
		0	0	_____	
		0	1	_____	
		1	0	_____	
		1	1	_____	

20. NOR

Symbol	Boolean	Truth table			Written expression
		A	B	output	
		0	0	_____	
		0	1	_____	
		1	0	_____	
		1	1	_____	